RESEARCH MEMORANDUM ON RURAL LIFE IN THE DEPRESSION

By DWIGHT SANDERSON

NYT

ARNO PRESS

A NEW YORK TIMES COMPANY

Reprint Edition 1972 by Arno Press Inc.

LC# 75-162844
ISBN 0-405-00847-3

Studies in the Social Aspects of the Depression
ISBN for complete set: 0-405-00840-6
See last pages of this volume for titles.

Manufactured in the United States of America

STUDIES
IN THE
SOCIAL ASPECTS
OF THE
DEPRESSION

Studies in the Social Aspects of the Depression

Advisory Editor: *ALEX BASKIN*

State University of New York at Stony Brook

Preface to the New Edition

FOR THE SMALL TOWNS, farm villages, and crossroad hamlets of America, the Depression started in 1921—the result of the failure of agriculture to readjust to post-World War I conditions. In some regions, traditional rural institutions broke down and families were compelled to abandon their homes and their farms. In other places, adversity became the binding force which brought and held divergent groups together. Under the banner of the Southern Tenant Farmers Union, black and white workers who once competed bitterly for jobs joined in common cause against their mutual foe, the landlord. In some instances, farmers engaged in direct action, blocking roads and withholding produce from the market until scarcity drove the prices of their goods upwards. The Farmers Holiday Association, undoubtedly influenced by trade union philosophy and techniques, led this movement in the normally conservative Corn Belt. The New Deal response to the farm crisis resulted in the establishment of the Agricultural Adjustment Administration, the Farm Credit Administration and the Resettlement Administration, as well as other government agencies. In this study, Professor Sanderson has recounted the role of governmental and voluntary organizations and their efforts to ameliorate and correct farm problems. He has described the droughts and the loss of valuable farm acreage and suggested what the consequences might be. In examining these issues he has also suggested areas where further research and investigation might be done. Trouble on the land in the 1930's was but one of the many signs of a nation in ill health.

Alex Baskin
Stony Brook, New York, 1971

BULLETIN 34

1937

RESEARCH MEMORANDUM ON RURAL LIFE IN THE DEPRESSION

By DWIGHT SANDERSON

*Chairman   Department of Rural Social
Organization—Cornell University*

PREPARED UNDER THE DIRECTION OF THE
COMMITTEE ON STUDIES IN SOCIAL
ASPECTS OF THE DEPRESSION

SOCIAL SCIENCE RESEARCH COUNCIL
230 PARK AVENUE   NEW YORK   NY

The Social Science Research Council was organized in 1923 and formally incorporated in 1924, composed of representatives chosen from the seven constituent societies and from time to time from related disciplines such as law, geography, psychiatry, medicine, and others. It is the purpose of the Council to plan, foster, promote, and develop research in the social field.

CONSTITUENT ORGANIZATIONS

American Anthropological Association

American Economic Association

American Historical Association

American Political Science Association

American Psychological Association

American Sociological Society

American Statistical Association

FOREWORD

*By the Committee on Studies in
Social Aspects of the Depression*

THIS monograph on research pertaining to rural life in the
depression is one of a series of thirteen sponsored by the
Social Science Research Council to stimulate the study of depression effects on various social institutions. The full list of titles is
on page ii.

The depression of the early 1930's was like the explosion of a
bomb dropped in the midst of society. All major social institutions, such as the government, family, church, and school, obviously were profoundly affected and repercussions were so far
reaching that scarcely any type of human activity was untouched.

It would be valuable to have assembled the vast record of
influence of this economic depression on society. Such a record
would constitute an especially important preparation for meeting
the shock of the next depression, if and when it comes. The facts
about the impact of the depression on social life have been only
partially recorded. Theories must be discussed and explored
now, if much of the information to test them is not to be lost
amid ephemeral sources.

The field is so broad that selection has been necessary. In
keeping with its mandate from the Social Science Research Council, the Committee sponsored no studies of an exclusively economic or political nature. The subjects chosen for inclusion were
limited in number by resources. The final selection was made by
the Committee from a much larger number of proposed subjects, on the basis of social importance and available personnel.

Although the monographs clearly reveal a uniformity of goal, they differ in the manner in which the various authors sought to attain that goal. This is a consequence of the Committee's belief that the promotion of research could best be served by not imposing rigid restrictions on the organization of materials by the contributors. It is felt that the encouraged freedom in approach and organization has resulted in the enrichment of the individual reports and of the series as a whole.

A common goal without rigidity in procedure was secured by requesting each author to examine critically the literature on the depression for the purpose of locating existing data and interpretations already reasonably well established, of discovering the more serious inadequacies in information, and of formulating research problems feasible for study. He was not expected to do this research himself. Nor was he expected to compile a full and systematically treated record of the depression as experienced in his field. Nevertheless, in indicating the new research which is needed, the writers found it necessary to report to some extent on what is known. These volumes actually contain much information on the social influences of the depression, in addition to their analyses of pressing research questions.

The undertaking was under the staff direction of Dr. Samuel A. Stouffer, who worked under the restrictions of a short time limit in order that prompt publication might be assured. He was assisted by Mr. Philip M. Hauser and Mr. A. J. Jaffe. The Committee wishes to express appreciation to the authors, who contributed their time and effort without remuneration, and to the many other individuals who generously lent aid and materials.

William F. Ogburn   Chairman
Shelby M. Harrison
Malcolm M. Willey

CONTENTS

Chapter I

Statement of the Problem, Objectives, and Method of Analysis

HOW has the depression affected the non-economic or social phases of rural life? Has the shock to our economic organization produced shifts in the social organization of country life which will be permanent? If so, what are these changes and what social research may be necessary to obtain a better knowledge of their significance from contemporary sources of information which may not be available later? If it is possible to recognize the significant tendencies in social change and to obtain sufficient knowledge concerning them, it may be possible to bring influence against those trends which are undesirable and to give encouragement to those which seem to be for rural welfare. These are the major objectives of the following inquiry.

In undertaking such an inquiry it is necessary that we first clearly state certain inevitable limitations which will tend to define its scope and to qualify the statement of the problem.

1. Although at the time of writing we seem to be coming out of the depression, we are still too close to it to gain a clear vision of its most important effects on rural society and whether they will be permanent or are but temporary. The critical changes produced by the depression will be clearly apparent only with the historical retrospect of another generation. Immediately after the World War we were flooded with literature purporting to show the changes which it had produced and the consequent necessity for "social reconstruction," but most of these aspirations soon disappeared and the world wagged on in much the

1

same old way, and even yet we are not quite sure of just what were the most significant social changes resulting from those four unprecedented years.

2. Inasmuch as the effects of the depression on rural life are the results of economic disorganization, it is difficult to separate the social consequences from the economic conditions producing them. The problems of human welfare and social organization arising on submarginal lands cannot be dissociated from those of land utilization and land economics. It is not within our province to consider the economic situation, but rather to attempt to indicate upon the basis of given economic conditions or tendencies what are the probable significant effects on the social aspects of rural life.

3. Coincident with the industrial depression have occurred two unprecedented droughts in 1933 and 1934, and in 1936, which have paralyzed some of the best agricultural regions, so that just what has been due to the economic depression and what has been due to drought it is impossible to determine.

4. Both conditions have led the federal government to establish new and unprecedented agencies for the assistance of agriculture through the Agricultural Adjustment Administration, the Farm Credit Administration, the Resettlement Administration, and other agencies, and to give federal funds for relief of the unemployed through the Federal Emergency Relief Administration (FERA) and its successor the Works Progress Administration (WPA). These activities of the federal government have mitigated and changed the effects of the depression on rural life, and the social situation which has resulted cannot be dissociated from them, for we cannot surmise what the effect of the depression might have been without them.

5. While the industrial depression starts with 1930 the depression in agriculture goes back to 1921 as an effect of lack of readjustment to postwar conditions, and although the agricul-

tural depression[1] became intensified after 1930, it is impossible to isolate the effects of the industrial depression of the '30's upon rural life without considering its impoverished condition in the '20's.

6. Finally, as in the case of the effects of a war, the depression in the main merely hastened and made acute certain tendencies which were already under way, and revealed undesirable social conditions which had long existed but which became so acute as to demand action. Unprecedented developments in transportation and communication and in the mechanization of agriculture; increased education; increased contacts with cities and the spread of urban ideologies, and other factors were responsible for an unusual degree of social change in rural life.[2] As a result it becomes a difficult problem to determine what has been the result of the depression and what would have occurred in "normal" times. As one correspondent states it, "We have, then, a problem of qualitative 'time-series analysis' in which it becomes impossible to segregate the variables." Our problem becomes, therefore, one of attempting to determine whether the depression has reversed or impeded any of the tendencies of social change or has so stimulated tendencies already under way as to hasten them to a crisis.

With the above factors so clearly delimiting the objectives of our quest, it is evident that the best we may hope to do is to point out certain conditions and tendencies in rural life in the United States, as they occur today, which are either definitely new or which have assumed such outstanding importance as to invite immediate investigation by the social scientist, without

---

[1] Hereafter we shall refer to the agricultural depression as the period 1921-1936, and the industrial depression as 1930-1936.

[2] Cf. *Recent Social Trends in the United States.* New York: McGraw-Hill Book Co. 1933. Vol. I. Chapter X, "Rural Life"; and Brunner, E. deS. and Kolb, J. H. *Rural Social Trends.* New York: McGraw-Hill Book Co. 1933

attempting definitely to determine whether they are the direct results of the industrial depression, or of the combination of depression, drought, and governmental policies.

This report is but a compilation of ideas and data obtained from many persons and sources. To obtain a preliminary orientation letters were written to about one hundred leaders in rural life throughout the country and very helpful replies were received from most of them. The author is particularly obligated to Dr. Edmund deS. Brunner for permission to read the manuscript of his report of the resurvey of agricultural villages[3] and to summarize his findings on several points. To him as well as to Dr. Carl C. Taylor, Dr. T. J. Woofter, Jr., and Dr. Carle C. Zimmerman the writer is indebted for reading the manuscript of the present monograph and for offering many helpful suggestions.

Before proceeding with our analysis it may be well to consider briefly some evidence of the extent of the depression as affecting agriculture, and to ascertain whether the effects of previous depressions on the social aspects of rural life may have any suggestions of phenomena which should be observed in the present situation.

EXTENT OF THE RURAL DEPRESSION

Although the effects of the industrial depression have been felt most keenly by the unemployed in the cities, the effect of low prices for farm products and losses due to drought over wide areas have caused unprecedented suffering in rural districts. From July 1933, when federal aid commenced, to October 1935, there were over 1,000,000 rural cases receiving relief from the federal government. The maximum was reached in January 1935, with 1,949,000 rural cases.[4] "These households contained a total

[3] Brunner, Edmund deS. and Lorge, Irving. *Rural Trends During Depression Years, 1930-1936.* New York: Columbia University Press. 1937

[4] Works Progress Administration, Division of Social Research. Research Bulletin. Series III, No. 1. August 26 1936. P. 3

of 8,600,000 persons, a number equal to 16 per cent of all rural persons in the United States in 1930."[5]

Another measure of the depression is the number of forced sales of farms. The Bureau of Agricultural Economics[6] gives annual estimates from 1930 to 1934 of the number of farms changing ownership by "forced sales and related defaults" per 1,000 total farms. These estimates for 12 month periods ending March 15 were as follows: 1930, 20.8; 1931, 26.1; 1932, 41.7; 1933, 54.1; 1934, 39.1. There is no means of determining how many of these farms may be included more than once in the five-year period, but if there were no duplications this would represent 18 per cent of all farms in the United States. However that may be, it is clear that over 5 per cent of all farms were subject to forced sale in 1933, and if they were of the average value of 1930, they would involve over three billion dollars' worth of farm property in that year. The highest number of forced sales was in the West North Central States in 1933, which rose to 72 per 1,000 farms. In this section the total forced changes in ownership for the five years 1930 to 1934 involved 23.4 per cent of all farms.

COMPARISON WITH PREVIOUS DEPRESSIONS

In attempting to assess the social effects of the depression one naturally reverts to the post-history of previous depressions to ascertain whether they will not furnish suggestions of what should be observed in the present situation. Unfortunately, we are practically balked in such an effort by the fact that historians have been mainly concerned with the political and economic significance of these crises. The sociological study of rural life has

---

[5] Federal Emergency Relief Administration. Division of Research, Statistics, and Finance. Research Bulletin H-4. October 29 1935. P. 2

[6] *Yearbook of Agriculture, 1935.* Washington, D.C.: United States Department of Agriculture. 1936. Table 477. P. 687

arisen in the present century and prior to that time but little attention had been given to its social institutions or social organization. Intensive historical research might reveal the social effects of previous depressions upon the social life of the countryside, but it is an open question whether the results would warrant the effort.

The one outstanding effect of previous depressions was that of emigration to new lands farther west, a theme which has been fully developed by contemporary historians from Turner to Adams.[7] After previous depressions there was always an accelerated movement to the virgin lands of the West, where the unfortunate could start life anew. The present depression marks the end of agricultural expansion in the United States. Not only is the good land fully occupied, but the depression has revealed that much poor land unadapted to agriculture must be relinquished from the plow. Furthermore, foreign markets have been lost and improved methods have made it possible to produce more than enough for our own markets on the better lands. Agricultural and economic opportunity now lies in the intensive development of specialties and in moving from poor to better land, both of which require capital.

The effect of this situation upon the psychology of the American farmer is difficult to determine, but it forms a challenge to social research. Heretofore there has always been the promise of a market on account of growing cities, but today we are confronted with the fact that our population will become static within two or three decades and that our cities are already failing to reproduce their populations. Although metropolitan areas will probably continue to grow for the next generation, it will be at a reduced rate, and their markets can be easily supplied

---

[7] *Cf.* Turner, Frederick J. *The Frontier in American History.* New York: Henry Holt & Co. 1920; Adams, James Truslow. *The Epic of America.* Boston: Little, Brown and Co., 1931

from the better lands by the use of better agricultural techniques. By and large, the American farmer is now forced to fight out his own salvation where he is, and without hope of materially better markets except as may result from a general rise in the standard of living or of better prices for his products, or except as he may, through his own efforts, obtain a larger share of the national income. Such a situation gives rise to a feeling of frustration which naturally finds vent in such direct action movements as the tobacco night-riders in Kentucky, the Farmers' Holiday Association, and milk strikes. These reactions have to do with the economic situation of the farmer, but have had no apparent effect upon rural social conditions. Whether a more dispassionate view of their situation, which will occur as the economic pressure is alleviated, will result in turning their attention to constructive efforts for the reconstruction of rural institutions so that they may enjoy a larger share of the benefits of modern culture, remains to be seen. To so analyze the present situation as to aid rural leaders in projecting a program to this end should be a major objective of social research.

The only permanent change in the organization of rural life caused by previous depressions is that resulting from the growth of the Grange (Patrons of Husbandry) in the early 1870's. Although organized in 1867 prior to the depression of 1873, its rapid growth to over three-quarters of a million members in 1874 to 1876, the fact that it grew most rapidly in the Middle West where the agricultural depression was most acute, and a study of the avowed objectives of the movement at that time, clearly show that its growth was directly related to the depression.[8] Although the chief objectives of the Grange have always

---

[8] *Cf.* Buck, Solon J. *The Granger Movement.* Cambridge: Harvard University Press. 1913. Although Buck does not directly discuss the effect of the depression upon the growth of the Grange it is implied in the facts he presents.

had to do with the economic side of agriculture, inasmuch as it is a farm family organization including men and women, it has increasingly become a social organization and has given its attention to the improvement of rural life. Its accomplishments in the latter field, although less tangible and less amenable to measurement, have probably been as significant in their effect on rural life as has been its effect on the technical and economic aspects of agriculture. The history of the Grange gives a clue for the investigation of the development of farmers' organizations during the present depression which will be considered later.

Although the Grange was the only permanent farmers' organization resulting from previous depressions, they gave rise to other movements which had notable influence. Thus, the famous Centralia Platform adopted by a convention of independent farmers at Centralia, Illinois in 1858 was the result of declining prices and the depression of 1857; the depression of the early '70's gave rise to the Farmers' Alliance; and from it grew the Populist Party as a result of the depression of the early '90's.[9]

METHOD OF ANALYSIS

To accomplish the objectives set forth in the introductory paragraphs the author first sought to determine what are the topics or problems which have most significance in revealing changes in rural life effected by the depression. These form the titles of the following chapters and their subheads. In each case there has been indicated the importance of the subject and how it has been affected or brought into prominence by the de-

---

[9] *Cf.* Wiest, Edward. *Agricultural Organization in the United States.* Chapter XVIII, "The Farmers' Alliance." Lexington: University of Kentucky. 1923

pression. These topics follow no logical scheme, but are rather an attempt to reveal the most important phases of rural life which have been strikingly influenced by the depression and which warrant research to determine the facts and to establish trends and tendencies which will have important influence on rural social organization in the future.

Secondly, the more important and suggestive research studies which have been made on each of these topics have been cited and briefly described to show what has been and may be done and to suggest methods of research which may be employed in future studies.

Lastly, the writer has indicated what research is needed to answer the questions raised and has suggested sources of data and research methods which seem applicable. In doing so the object has been not so much to outline definite methods of research or project statements, as to call attention to research needed and to suggest possible methods. For suggestions as to research methods the student may obtain assistance from the series of bulletins on *Scope and Method* published by the Social Science Research Council under the direction of the Advisory Committee on Social and Economic Research in Agriculture, edited by John D. Black. This series consists of 21 bulletins, a few of which are noted below as being particularly relevant to the subject under discussion.[10] After all, one who is

[10] Melvin, Bruce L. *Research in Rural Population.* Bulletin 4. May 1932. Pp. 149; Brown, Josephine C. *Research in Rural Social Work.* Bulletin 5. July 1932. Pp. 106; Zimmerman, C. C. *Research in Farm Family Living.* Bulletin 11. April 1933. Pp. 209; Zimmerman, C. C. *Research in Rural Organization.* Bulletin 12. March 1933. Pp. 160; *Research in Farm Labor.* Bulletin 16. June 1933. Pp. 84; Anderson, C. Arnold and Smith, T. Lynn. *Research in Social Psychology of Rural Life.* Bulletin 17. June 1933. Pp. 130; Foster, Robert G. and Hamilton, C. Horace. *Research in Rural Institutions.* Bulletin 18. June 1933. Pp. 112. New York: Social Science Research Council

competent to carry on or direct research in this field must have the necessary knowledge of materials and methods and have sufficient imagination and inventiveness to apply them to the problem under investigation. Our task is rather to stimulate further analysis of the topics discussed, and to indicate what seem to be feasible methods of investigation.

To avoid repetition and to preserve unity these three phases of our discussion have usually been considered together under each topic.

Chapter II

Changes in Rural Population, Its Composition and Movement

INASMUCH as the number and composition of the population of a given area is a primary factor conditioning its social
life, significant population changes which may have been caused
by the depression would be of primary importance in evaluating
depression effects on rural society.

Population changes may occur by natural increase or decrease
and by immigration and emigration. Each of these topics will
be considered with regard to what has happened during the depression.

With respect to the birth rate Thompson and Whelpton[1] have
shown that the standardized birth rate for the rural United States
declined much more rapidly between 1920 and 1929 than did
the urban rate or that of the United States as a whole, and this
decline was most pronounced in five southern states. Brunner and
Kolb[2] state that in 347 counties scattered over the United States
the fertility ratio, the ratio of children under 10 to women 20
to 45 years of age, declined more rapidly between 1920 and
1930 than in the previous decade. The counties studied were
arranged in four concentric tiers around eighteen cities, and it

---

[1] Thompson, W. S. and Whelpton, P. K. *Population Trends in the United
States*. Fig. 32, p. 273. New York: McGraw-Hill Book Co. 1933
[2] Brunner, E. deS. and Kolb, J. H. *Rural Social Trends*. Table 52, p. 126.
New York: McGraw-Hill Book Co. 1933. In this table the sections on total population and rural population have obviously been transposed. The rural section has,
therefore, been reprinted with the changes between the decades computed from it.

11

was found that the fertility ratio of the outer two tiers declined more rapidly in both decades than did the inner two tiers, those nearer the cities, so that while in 1920 the ratio of the outer tiers was much higher, in 1930 it was practically the same as that of the inner tiers. This is shown in Table I.

TABLE I

NUMBER OF CHILDREN UNDER 10 PER 1,000 WOMEN 20 TO 45 YEARS OF AGE IN
THE RURAL POPULATION OF 347 COUNTIES, BY TIERS OF COUNTIES IN
18 AREAS, AND PERCENTAGE DECLINE BETWEEN DECADES[a]

YEAR	CITY COUNTY	TIERS OF COUNTIES (ARRANGED IN ORDER OF PROXIMITY TO CITY)			
		1	2	3	4
1910	1,381	1,443	1,482	1,522	1,635
1920	1,309	1,392	1,440	1,470	1,509
1930	1,234	1,318	1,363	1,374	1,345
		Percentage decline			
1910 to 1920	5.2	3.5	2.8	3.4	7.7
1920 to 1930	5.7	5.3	5.3	6.5	10.9

[a] Brunner, E. deS. and Kolb, J. H. *Rural Social Trends.* Table 52. P. 126. New York: McGraw-Hill Book Co. 1933

As the authors state, "Apparently the fecundity rate of the population is declining more rapidly in tiers more remote from, than in tiers nearer to cities; which suggests that the decline began later in the outlying tiers than in the cities and counties close to the cities."

The validity of this conclusion seems to be questionable and to demand further study, for an attempt made by the Scripps Foundation for Research in Population Problems[3] to corroborate the results of Brunner and Kolb resulted in conflicting evidence and indicates that other factors are involved. However, it may be noted that this more rapid decline in fertility in the more

[3] Whelpton, P. K. "Geographic and Economic Differentials in Fertility." *Annals of the American Academy of Political and Social Science.* 188:43, November 1936

rural areas is rather confirmed by a regional analysis of birth rates in Kansas made by Clark and Roberts.[4]

It is well known that the birth rate declines during business depressions.[5] It seems possible that the larger decline in the rural birth rate during the '20's, both as compared with the city rate and the rural rate of the previous decade, may have been partly due to the agricultural depression of that decade, and raises the question as to whether this tendency may have been accelerated by the industrial depression of the '30's. If there has been an acceleration of the previous decline in the rural birth rate, this might be of considerable importance, particularly in the more rural sections; for although it is probable that with a revival of business and a better income the rate would tend to rise, it might be that the attitudes toward reproduction developed during the depression might carry over and have a lasting effect in reducing the rural birth rate more rapidly than would have occurred otherwise. Local studies[6] of the birth rate statistics of the last five years are therefore desirable for they may show that

[4] Clark, Carroll D. and Roberts, Roy L. *People of Kansas.* Topeka: Kansas State Planning Board. October 1936. P. 132. See also Nelson, Lowry, and Hettig, T. David. "Some Changes in the Population of Utah as Indicated by the Annual L.D.S. Church Census, 1929-1933." *Utah Academy of Sciences, Arts and Letters.* 12:107-118. 1935

[5] *Cf.* Thomas, Dorothy. *Social Aspects of the Business Cycle.* London: G. Routledge & Sons, Ltd. 1925; Lorimer, Frank and Osborn, Frederick. *Dynamics of Population.* P. 292. New York: The Macmillan Co. 1934

[6] For example see Lively, C. E. and Folse, C. L. *The Trend of Births, Deaths, Natural Increase and Migration in the Rural Population of Ohio.* Columbus: Ohio State University. Department of Rural Economics and the Ohio Agricultural Experiment Station Bulletin 87. April 1937. Mimeo.; also Hamilton, C. Horace and York, Marguerite. "Trends in the Fecundity of Married Women of Different Social Groups in Certain Rural Areas of North Carolina." *Rural Sociology.* 2:192-203. June 1937. They have shown the possibility of obtaining the fertility rate from survey data. They find a decline from 1915 to 1931, but that this was checked in 1932 to 1934. The general effects of the depression on fertility are discussed by Stouffer, Samuel A. and Lazarsfeld, Paul F. in *Research Memorandum on the Family in the Depression.* Chapter V. (monograph in this series)

in many sections, because of the declining birth rate, the assumed future surplus of rural population which might result from less migration to cities may not be so great.

In this connection it would be desirable to investigate the birth rate of rural families receiving relief during the depression with those of non-relief families of approximately the same status in the same communities.[7] The only study of the birth rate of rural relief families we have seen is the one by Hamilton and York (*loc. cit.*) of certain rural areas in North Carolina. They found no significant difference in the trend of the fertility rate between relief and non-relief women during the depression. It will also be desirable to have studies of the birth rate of farm families in relation to their income. Notestein[8] has shown the differences in fertility between wives of farm owners, farm renters and farm laborers in the North Central States for 1900 and 1910, and this has been confirmed by Hamilton and York for North Carolina, but no studies have been made to show the differences in fertility according to income within the farm owner or farm renter classes. This could be done best by the survey method.

Partially offsetting the declining birth rate has been the decrease in net migration from rural to urban places.[9] Estimates of the U. S. Bureau of Agricultural Economics[10] show that in 1932

---

[7] Studies of the difference of fertility of families on relief in cities have been made by Stouffer, Samuel A. "Fertility of Families on Relief." *Journal of the American Statistical Association.* 29:295-300. September 1934; and the general situation has been discussed by Notestein, Frank W. "The Fertility of Populations Supported by Public Relief." *The Milbank Memorial Fund Quarterly.* 14:37-49. January 1936

[8] Notestein, Frank W. "The Differential Rate of Increase among the Social Classes of the American Population." *Social Forces.* 12:19-33. October 1933

[9] For a detailed critique of the statistics of internal migration see Thompson, Warren S. *Research Memorandum on Internal Migration in the Depression.* Chapter III (monograph in this series)

[10] Revised estimates in "Farm Population Estimates, January 1 1936." Washington, D.C.: U. S. Department of Agriculture. Bureau of Agricultural Economics. Press release of October 27 1936

there was a net gain to the rural population caused by an excess of the urban-rural over the rural-urban migration, for the first time since the estimates of farm-city migration began in 1920. "In the years of prosperity, before 1929, more people moved away from farms than moved to farms and by 1930 the number of people living on farms was less than at any time since 1910. Since the depression began, fewer people have moved to towns and cities and some people have moved back to farms, and today there are more people on farms than there were when the depression began; nevertheless, the number of persons on farms today is less than it was at the beginning of 1910."[11]

A comparison of the figures for 1930-1934 with those for the five years preceding 1930 shows immediately that there was much less migration to and from farms during the depression years. The number of persons leaving farms decreased from 10¾ million to 7 million; the number of persons moving to farms decreased more than 1 million, from 7¾ to 6½ million. Since the number leaving farms decreased more than the number moving to farms, there was also a sharp reduction in the net losses as a result of the movement from farms to villages, towns, and cities. During the 5 predepression years, the number of persons who left farms was nearly 3 million greater than the number who came to farms, an average of 600,000 persons each year. During the depression years this large annual outflow of farm people was slowed down to such an extent that the total number for the entire 5-year period was only 600,000 or about as much as the annual average for the five years just preceding 1930. Although many people left towns and cities to move to farms after 1930, the growth of the farm population was even more affected by the fact that fewer persons left farms than would have been the case had the conditions of the late twenties continued. . . .

If movements to and from farms only are considered, there was a net loss in farm population during 1930, 1931, 1933, and 1934, though in each case these losses were less than any reported since 1921. Only during 1932 were more persons reported as moving to farms than moved from farms, although during 1931 the two movements nearly balanced. But by 1933 the loss due to migration was already greater than in 1930 and it increased still further during 1934 and 1935.[12]

---

[11] *Ibid.*

[12] "Movement to Farms." *U. S. Census of Agriculture, 1935.* Washington, D.C.: U. S. Department of Commerce. U. S. Bureau of the Census. Press release. June 13 1936. Mimeo.

The agricultural census of 1935 shows nearly two million persons on farms who lived on a non-farm residence five years earlier. This number constituted 6.3 per cent of the total farm population in 1935. Some of the increase reported by the census is undoubtedly due to a more zealous enumeration of farms than in 1930 and also to families who commenced to farm small places on which they had been living previous to 1930, particularly near cities.[13] Although the movement to farms may be somewhat discounted by changes in enumeration, this does not invalidate the evidence of a movement to rural territory. This movement to farms by non-farmers has occurred mostly around cities and in some of the poorer agricultural sections, as has been well summarized by the Bureau of the Census.

The movement to farms has been most extensive in five regions. Perhaps the largest of the regions comprises the Appalachian subsistence farming area extending along the hills and mountains from Pennsylvania to Alabama. Unemployed miners, lumber workers, factory employees, and others who were unable to find work in industrial pursuits, totalling in all between a third and a half a million persons, returned to the small hill and valley farms which were once abandoned, or to farms being operated by their relatives or friends. Areas around the industrial centers of New England, New York, Michigan and Ohio furnish typical examples of other regions where there has been a large influx of people to farms. Between a third and a half million persons from these industrial centers have resettled on idle farms or have engaged in part-time farming in the surrounding farm areas. The three regions in which the movement to farms was extensive included the cutover lands of northeastern Minnesota and northwestern Wisconsin, the Ozark Mountain and eastern Oklahoma area and the Pacific Coast valleys. Into each of these three areas over 100,000 persons have moved from cities, towns, villages, or other non-farm residence.

Most of this flow of persons back to the farms represents the countryward migration of the unemployed and others unable to obtain work in factories, mines, lumber mills, etc., to escape idleness, reduce expenses, and to raise food for family use. The average of 3 persons per farm reporting this movement, and the increase of over half a million

---

[13] See Thompson, Warren S. *Research Memorandum on Internal Migration in the Depression*

farms since 1930, suggest that this has been largely a movement of whole families rather than the migration of single individuals. These families have returned to farms once abandoned, to new farms, and to unoccupied houses on farms operated by their relatives and friends.

Not all of these newcomers have been unemployed nor have all remained unemployed. In many areas, thousands have engaged in part-time farming and have supplemented their urban, or industrial, income with rural living. Improved roads, reduced transportation charges, and cheaper living in the country than in the city have brought thousands of families back to the land to live and have been factors in holding them there.[14]

The geographical distribution movement may be clearly seen from spot maps, prepared by the Bureau of Agricultural Economics, of the increase in the number of farms from 1930 to 1935, and of the number of people living on a farm January 1, 1935 who were not living on a farm five years before.[15] Between 1920 and 1930 the largest rate of increase in the rural population was in the increase of the rural non-farm population near cities. This increase was undoubtedly accelerated by the depression, but no census data concerning it will be available until 1940, except for state censuses such as that in Michigan in 1935. Estimates of the movement of the total rural population have, however, been made by Goodrich[16] and Allin[17] based on the school censuses of various states which show that the general

[14] "Movement to Farms," *U. S. Census of Agriculture, 1935*. U. S. Bureau of the Census. Press release. June 13 1936

[15] Goodrich, Carter *et al. Migration and Economic Opportunity*. Philadelphia: University of Pennsylvania Press. 1936. Fig. 75, p. 514; and Baker, O. E. "Rural and Urban Distribution of the Population in the United States." *Annals of the American Academy of Political and Social Science*. 188:267. Fig. 6. November 1936

[16] Goodrich, Carter, *et al. Op. cit.* Pp. 73, 190, 508; Goodrich, Carter, Allin, Bushrod W., and Hayes, Marion. *Migration and Planes of Living 1920-1934* Bulletin 2, Study of Population Redistribution. Philadelphia: University of Pennsylvania Press. 1935

[17] Allin, Bushrod W. and Parsons, Kenneth H. "Changes in the School Census Since 1920." *Land Policy Review*. I. Sup. 1. June 1935. Washington: Agricultural Adjustment Administration. Mimeo.

movement to farms described by the census has been true for the rural population as a whole. In addition to showing that the most rural counties in many states increased in school population from 1929 to 1933, Allin and Parsons give evidence that the gains in population during these years were highest in the counties with the poorest land.

For many states the division of agricultural counties into quartiles on the basis of income per person in 1929 classified the counties broadly according to land quality. To the extent that the income index does measure land quality, another conclusion of this study is that poor land areas lost larger proportions of their population than good land areas during the period of industrial prosperity from 1922 to 1929, and gained larger proportions during the depression. Moreover, the poorer the land, the greater were the recent increases in population. This is especially true of land areas suitable for subsistence farming, and not too remotely located with reference to urban and non-farm centers of population.[18]

This general trend for population to increase most on the poorest land is confirmed by observations of sociologists and economists in Tennessee, Kentucky, South Carolina, Louisiana, Arkansas, and Oklahoma.[19] However, in several of these states the increase of population on poorer land was most in industrial or mining counties in which there was a large natural increase, emigration had been checked, and local industrial workers had gone to farming as a means of subsistence, rather than from immigration. A similar trend has taken place in Ohio as shown by censuses of mobility made by Dr. C. E. Lively[20] in 10 townships. He finds that, from 1920 to 1930, the net loss by migration was largest in the poorer southeastern section of the

---

[18] Ibid.

[19] Cf. McMillan, Robert T. "Some Observations on Oklahoma Population Movements Since 1930." Rural Sociology. 1:332-343. September 1936; and Thompson, Warren S. Research Memorandum on Internal Migration in the Depression

[20] Lively, C. E. and Foott, Frances. Population Mobility in Selected Areas of Rural Ohio, 1928-1935. Wooster: Ohio Agricultural Experiment Station. Bulletin 582. June 1937

state, but that from 1930 to 1935 there was less net loss in this section than in the better farming section of western Ohio. This was due to the fact that outward migration decreased much more in the southeastern than in the western section, in which it was nearly the same as in the previous decade.

More important than immigration in accounting for the increase of rural population during the depression has been the partial stoppage of emigration to towns and cities. For many years there has been a steady rural-urban migration, particularly of young people, which is at once apparent in any study of the age distribution of urban and rural populations as given by the census. During the decade 1920-29 this migration averaged 669,000 persons per year according to the estimates of the Bureau of Agricultural Economics.[21] With no employment available in the cities this migration decreased and the movement was reversed for the year 1932. Rural-urban migration was greatly reduced in 1933 and 1934, being only about one-half the movement from farms during the decade 1920-29. Obviously, as explained above in the quotations from the estimates of the Bureau of Agricultural Economics, this has resulted in a considerable increase in rural population in those areas from which rural-urban migration had previously been heaviest. Furthermore, this stoppage has most seriously affected young people 20 to 30 years of age, who formed the bulk of the rural-urban migration. Goodrich[22] estimates that, of the total net migration between 1920 and 1930, 50 per cent was of persons between ten and twenty years of age in 1920, and that over 40 per cent

[21] *Yearbook of Agriculture, 1935.* Washington, D.C.: U. S. Department of Agriculture. P. 674. On p. 690, the total net migration is estimated at 8,000,000 or 800,000 per year for this decade, which estimate is probably much too large.

[22] *Op. cit.* P. 690. This is essentially confirmed by Table II of Dorn, Harold F. and Lorimer, Frank. "Migration, Reproduction, and Population Adjustment." *Annals of the American Academy of Political and Social Science.* 188:280-289. November 1936. They also give an excellent summary of the increase of farm population caused by the decrease of emigration.

of those on farms in 1920 who were between ten and twenty years of age moved into cities during the ensuing decade. This is confirmed by the detailed study of migration in North Carolina made by Hamilton[23] who states:

Nearly three-fourths (73.1 per cent) of all the net migration loss is accounted for by the age groups from 15 to 35 years of age; and approximately one-half (50.8 per cent) is accounted for by the age groups from 20 to 30 years. About one-third of the expected population in 1930 from 20 to 30 years of age were lost to the farms by net migration.

No measurement of this stoppage of migration of rural youth from 1930 to 1935 is possible by the methods used by the authors cited until the 1940 census is available and the situation may then be obscured by a subsequent resumption of migration in the latter half of the decade. However, this stoppage of migration of youth may be determined by surveys, as has been demonstrated by Hamilton,[24] who shows that in North Carolina "young people from families on relief in 1934 left their parental homes at a higher rate during a period of relative prosperity, but that in periods of relative depression they left their homes at a lower rate than did young people from non-relief families."

It should be clear that the increase of farm population between 1930 and 1935 was due to a relative decrease in the rate of farm-urban migration which was larger than the decrease in the urban-farm migration. Total migration decreased during this period, but the farm-urban movement was slowed up more than the movement to farms, so that a net increase in farm population resulted. On the other hand, although this increase was chiefly due to a decrease in farm-urban migration, this migration was still so large that there would have been a net loss of farm population had there been no migration to farms from cities.

---

[23] Hamilton, C. Horace. *Rural-Urban Migration in North Carolina 1920 to 1930.* Raleigh: North Carolina Agricultural Experiment Station. Bulletin 295. February 1934. P. 38

[24] Hamilton, C. Horace. "The Annual Rate of Departure of Rural Youths from Their Parental Homes." *Rural Sociology.* 1:164-179. June 1936

Studies of rural mobility have recently been conducted in Ohio, Iowa, North Dakota, South Dakota, Arizona, Maryland, Kentucky, and North Carolina. These, using a uniform schedule, have been made in cooperation with the Federal Emergency Relief Administration and the Bureau of Agricultural Economics and will undoubtedly make important contributions to our knowledge of these matters, as they include all open-country and village (under 2,500) populations of the areas studied. These studies were inaugurated by Dr. C. E. Lively of Ohio State University, who has had general supervision of their tabulation. A report[25] on the results obtained in Ohio—continuing the previous studies[26] on mobility made by him in that state—indicates that migration to the villages has been considerably heavier than migration to the open country, and that failure of adult children to migrate from the rural districts has been of at least twice the importance as the return migration from the cities as a cause of increased population. This report very clearly shows the advantages of a special mobility census of this sort for definitely locating the movements of rural population and the desirability of machinery for quickly obtaining a widespread sample in periods of shifting movements. Repeated studies of this sort conducted in the same areas at regular intervals will be highly desirable to give exact information as to population mobility, and as a means of interpreting estimates of mobility as made by the U. S. Bureau of Agricultural Economics.

The Pacific Coast states had a considerable immigration from the drought areas of the Rocky Mountain states and western plains. A recent study of the rural immigrants to Washington[27]

---

[25] Lively, C. E. and Foott, Frances. *Population Mobility in Selected Areas of Rural Ohio, 1928-1935.* Wooster: Ohio Agricultural Experiment Station. Bulletin 582. June 1937

[26] Lively, C. E. and Beck, P. G. *Movement of Open-Country Population in Ohio.* Wooster: Ohio Agricultural Erperiment Station. Bulletin 467. November 1930; Bulletin 489. September 1931

[27] Landis, Paul H. *Rural Immigrants to Washington State, 1932-1936.*

is of interest in showing the possibility of making a study of such movements through rural school teachers.

Important light on the mobility of farm operators may also be obtained by analysis of the period of farm occupancy as given by the Census of Agriculture of 1935, which compares the period of occupancy for owner and tenant operators with that in 1930 for six classes of length of occupancy. If these data can be analyzed within states by counties, and for agricultural regions, it will undoubtedly reveal the types of areas and the types of farms most affected by mobility. Special tabulations of occupancy by size of farm and for part time farms would also be of value. In reply to an inquiry concerning the availability of these data, the chief statistician for agriculture of the U. S. Census Bureau, Mr. Z. R. Pettet, writes: "We are pleased to inform you that county table No. 4 of the second series 1935 Census of Agriculture bulletins contains the data on years of farm occupancy by counties. Similar data for 1930 were recorded upon intermediate sheets, but not published by counties. The information could be secured by paying for the cost of photostating the sheets or copying them. Special tabulations could be made by showing this material by size of farms and part-time farmers. It would require a special run to obtain this material in this form."

One factor in the migration to and from farms, and of the rural non-farm population, which has not been sufficiently studied is the movement from the open country to villages and small towns. Beck and Forster,[28] in their study of *Rural Problem Areas,* found that in the Appalachian-Ozark area there was little movement of relief cases between 1930 and 1934, and in the Lake States Cut-over area "10 per cent of the open country families receiving relief had moved there from towns and villages, and

---

Pullman: State College of Washington Agricultural Experiment Station. Rural Sociology Series in Population. No. 2. July 1936. P. 21. Mimeo.

[28] Beck, P. G. and Forster, M. C. *Six Rural Problem Areas.* Research Monograph I. Washington, D.C.: Federal Emergency Relief Administration. Division of Research, Statistics, and Finance. 1935. Pp. 66-67

11 per cent from cities since 1930."[29] In the four other areas, involving the Short Grass and Cotton areas, there was a reverse movement from the open country to villages and towns of under 5,000 due to the unemployment of farm tenants and farm laborers. Further evidence on this point will be available from the mobility studies made by surveys in the eight states mentioned above. The study made in Ohio suggests to the authors[30] that "during the movements of population in 1930-35 the villages became an area of concentration, probably receiving population from both cities and open country." Data will also be available from the resurvey of 140 villages throughout the country made by Dr. E. deS. Brunner in cooperation with the Bureau of Agricultural Economics. A detailed study of school censuses for 1930 to 1935 of villages and towns with those of the surrounding country districts in those states in which such a division is available, and where village and open-country residence is recorded, would seem to be worthwhile to throw light on this point, particularly in states which show a net loss in farm population from 1930 to 1935.

This country-village movement is of considerable importance if it means that during the depression a large proportion of rural families needing public relief gravitated to the villages. Upon this Beck and Forster furnished rather definite evidence: "The unemployed relief clients tended to migrate into, or remain in the towns and villages." . . . "In all except the Appalachian-Ozark Area the percentage of the unemployed living in villages and towns was considerably greater than for the unemployed, among male family heads usually engaged in agriculture."[31]

---

[29] A study of the relief population in six counties in Oregon shows that the major movement was to the farms. See Hoffman, C. S. *Mobility and Migration of Rural Relief Households in Six Oregon Counties*. Corvallis: Oregon State Agricultural College. Agricultural Experiment Station. Circular of Information No. 155. June 1936. Pp. 8. Mimeo

[30] Lively, C. E. and Foott, Frances. *Op. cit.*

[31] *Op. cit.* P. 69, and figure 11

Lively and Foott[32] also note this tendency in Ohio where "In the open country only 12.9 per cent of the households was classified as relief cases; whereas in the villages the intensity of relief rose to 19.7 per cent of all households."

The obvious effect of the depression on the usual migration and the evident need for giving assistance for migration from "stranded areas," have given unprecedented interest to the whole subject of rural migration. Although the need of a solution for the problems of overpopulation of some rural sections is evident, the more the problem is studied the more difficult any policy of encouraging large scale migration by governmental agencies is seen to be.[33] The situation has therefore shown the desirability of obtaining as much information as possible concerning all phases of rural migration, social as well as economic.

One matter upon which more accurate information is needed with regard to different regions and sections is the places to which migration occurs, particularly with regard to young people leaving home. In a western New York county, Dr. W. A. Anderson[34] has shown that over half of the young people migrate to some place in the same county and that three-fourths migrate to the same or an adjoining county; yet it is apparent that from the southern Appalachian area there is a large migration involving considerable distances to northern cities. It would be helpful to have accurate data with regard to this matter for many sections.

It also seems evident that during a depression there is a large return migration to the poorest sections from which outward migration has previously been heavy because of overpopula-

---

[32] Op. cit. P. 7

[33] Goodrich, Carter et al. Migration and Economic Opportunity. Philadelphia: University of Pennsylvania Press. 1936. Ch. XII

[34] Anderson, W. A. Mobility of Rural Families. II. Ithaca: Cornell University Agricultural Experiment Station. Bulletin 623. March 1935. P. 20

tion.[35] Thus the poorest sections have the largest fluctuations caused by migration and during a depression are forced to bear the burden of an increased population although they are least able to support it. If this be true, it would seem to warrant a policy of federal assistance to these areas in time of depression. The facts as to what extent return migration actually occurred during the depression, and the type of people who were the return migrants, should be more clearly established by further surveys.

In this connection it would be desirable to ascertain the reasons for returning to the poorer counties from which the migrants presumably originated. Is this due to the fact that they desire to get back to the home county where they are acquainted, and where they feel there is probably a possibility of eking out a more comfortable existence because of their knowledge of the situation and the possible help of relatives and friends?

A related question is the effect of return migration upon the home community. It has been observed that in some counties the return migrants who have been engaged in industrial employment elsewhere have brought back entirely new ideas and standards of living and have tended to stir up dissatisfaction in the home community.[36] If repeated oscillations of migration occur, this might be a considerable factor influencing social change in the communities affected.

The old question of the selectivity of migration also arises in a new form. There has long been a contention that those migrating from rural communities to towns and cities are the more ambitious and enterprising and that, therefore, the less able individuals are left behind, and rural degeneration results. So far as we are aware, there has never been any very clear cut

---

[35] Cf. Goodrich, et. al. Op. cit. P. 512

[36] See Thompson, Warren·S. Research Memorandum on Internal Migration in the Depression. Chapter II

evidence that this is the case, but it is a natural and popular logical assumption. In the case of return migrants there is the presumption that it is the less able persons who have lost their jobs in the cities and have therefore returned to the country. This is a difficult problem, but warrants much more comprehensive and exact research than has ever been given it.

Finally, the question arises whether, however much we might know about the facts of migration, the overpopulation of such sections as the southern Appalachian highlands and parts of the Cotton Belt can be permanently alleviated by any process of migration unless it can be indefinitely maintained. With the present outlook for increased urban employment this seems improbable. Is not the solution rather in finding means for reducing the birth rate? To this end it might be worthwhile to ascertain those counties within an overpopulated area which have the lowest birth rates and which have shown the most rapid decline in birth rate, and then, by survey methods, attempt to ascertain what factors seem to have been responsible for lowering the fertility rate.

Should the stoppage of rural-urban migration prove somewhat permanent it will have a most serious effect on the poorest agricultural sections where the rate of reproduction is highest and where the possibility of existence is least favorable. For such areas it would mean either a lower standard of living, already meager, continued relief, or emigration. These changes in population increase or decrease have thus focused attention upon the critical problem of those living on marginal lands, and have made it a major social problem similar to that revealed by the discovery of the plight of the urban slums in the last decades of the nineteenth century.

The revelation of the problem of excessive population on submarginal lands, such as the southern Appalachian highlands, raises questions of public policy with regard to such areas, whose ultimate solution will depend quite largely upon the national attitude toward basic human values; but a large amount of research by social scientists will be necessary before the situation can be so clearly defined as to make wise decisions possible.

Would the people in these areas be better off or happier if they were encouraged to move to better areas by means of government aid? Is it desirable to maintain a reservoir of cheaper labor for the industrial centers? If state and government aid were extended to these poorer regions for the maintenance of schools, roads, and other facilities would it be possible to maintain a reasonably satisfactory type of civilization, or would such aid only encourage the present high rates of reproduction? If better facilities for education and communication were furnished these regions would there perhaps result a higher standard of living, which would automatically produce a decreased birth rate and more emigration? Would the cost of such a program be excessive? Would it be better to abandon considerable areas of the poorer lands and return them to forest or grazing? These are some questions which probably cannot be solved for whole regions, and which can be answered only after careful studies of the conditions in restricted areas, as related to conditions throughout the entire country, are made by agriculturists, economists, and sociologists.

To summarize: With respect to the relation of the depression to problems of rural population, it has been shown that the striking changes which have occurred, together with the problems of encouraging migration from some areas, have greatly stimulated interest which should lead to research along several lines. The most important of these are: (1) the changes, by counties, in fertility ratios as related to various factors, and the special study of those counties in overpopulated areas which have shown the highest rate of decline; (2) the birth rate of rural families on relief; (3) measurements by means of surveys of the stoppage of migration of rural youth during this period; (4) the importance of recurring surveys of rural mobility for identical sample areas; (5) the analysis, by counties, of the period of occupancy of farms as related to various factors affecting it; (6) the effect of unusual immigration on the rural community.

Chapter III

Social Corollaries of Agricultural Readjustment Problems

THE depression forced upon public attention the fact that important readjustments in types and methods of agriculture must be made to meet conditions which had been changing for some time, but which it brought to a crisis. These problems of agricultural readjustment involve important consequences to the social aspects of rural life which require investigation. Some of the more important phases of this process of agricultural readjustment are: (1) the new importance of regionalism, (2) the consideration of public policy with regard to the use of submarginal and marginal lands, (3) the increase of part time farming, especially near cities, and (4) the questions arising in policies of rehabilitation and resettlement. The social aspects of each of these topics and the research necessary concerning them will be considered in this chapter.

REGIONALISM

With the attempt to devise ways and means for the readjustment of agriculture it at once became apparent that no one formula would be applicable to the whole country, that the problems of the Cotton Belt were different from those of the Northeast, etc. As studies of land utilization,[1] soil conserva-

---

[1] Beck, P. G. and Forster, M. C. *Six Rural Problem Areas.* Research Monograph I. Washington: Federal Emergency Relief Administration. Division of Research, Statistics, and Finance. 1935

tion, and agricultural readjustment proceeded, more and more emphasis was placed upon the delimitation and mapping of agricultural regions and the description and analysis of their agricultural assets and liabilities.

When the problems of human relief became more acute and the federal government extended aid to the states through the FERA, it soon became apparent that certain rural areas were in much worse condition than others. An analysis of the rural relief situation revealed six problem areas which were made the subject of intensive studies and a comprehensive report.[2] This report definitely outlined four or five cultural regions in which the underlying economic factors responsible for the large relief lists and the concomitant social conditions and problems were essentially different. In both the Appalachian-Ozark and Lake States Cut-over areas, the fundamental economic disorgani- zation was the aftermath of the exploitation of natural resources with no consideration of its ultimate consequences to human welfare. This was complicated in the first area by overpopula- tion and in the second by too rapid settlement on poor land. In the two areas of the Cotton Belt the major problems were the result of the gradual breakdown of the sharecropper and "furnishing" system as a result of which the patriarchal system of caring for dependents shifted to one of public relief. In- adequate education, however, was also a factor. In the Short- grass areas of the western great plains the swelled relief rolls were the immediate effect of drought, but droughts are char- acteristic of these areas, and the intensity of the destitution was intimately related to speculative farming with high investments for machinery. Except for these areas the problems of rural relief were not acute, at least for farm operators, but on the Pacific Coast the conditions of transient farm labor became acute and are still unsolved.

Thus the relief situation brought into focus the fundamental

[2] *Ibid.*

differences among the various regions in economic resources and possibilities and accompanying social cultures.

As huge sums were made available for public works projects as a means for furnishing private employment and stimulating industry, and as large sums were devoted to work relief upon local public projects, it was but logical to give consideration to national and state planning on a broader basis than previously. The National Resources Board was created and stimulated the organization of state planning boards in most of the states. As a result of its studies the National Resources Board (now the National Resources Committee) saw the necessity of dealing with planning problems by regions and has brought out a comprehensive report on Regional Factors in National Planning.[3]

The creation of the Tennessee Valley Authority with its extensive program of hydroelectric development for a whole region also focused attention on problems of regional planning, and the TVA has accumulated a large amount of data on that region.

The economic and political issues which have come to the front during the depression have brought into focus the conflict of interests between major regions of the United States and have confirmed the wisdom of the observation of Frederick J. Turner made many years ago.

Indeed, the United States is, in size and natural resources, an empire, a collection of potential nations, rather than a single nation. It is comparable in area to Europe. . . . Within this vast empire, there are geographic provinces, separate in physical conditions, into which American colonization has flowed, and in each of which a special society has developed, with an economic, political, and social life of its own. Each of these provinces, or sections, has developed its own leaders, who in the public life of the nation have voiced the needs of their section, contending with the representatives of other sections. . . . In short the real federal aspect of the nation, if we penetrate beneath the constitutional forms to the deeper currents of social, economic and political life, will

[3] National Resources Committee. *Regional Factors in National Planning and Development.* Washington: Government Printing Office. 1935

be found to be in the relations of the sections and nation, rather than in the relation of states and nation.[4]

As a result of this general movement toward regional analysis, which has been stimulated by the depression, we have already had two excellent monographs of major regions, and another of the western Great Plains will probably result from investigations now being made of drought conditions. The first of these,[5] based on original investigation as well as documentary sources, is the first example of an extensive piece of research conducted cooperatively by a department of the federal government and four of the state agricultural experiment stations with several private national agencies. The second,[6] supported by private funds, is an equally impressive compilation and analysis of statistical material and documentary evidence made by a corps of university men under the auspices of a committee of the Social Science Research Council. Similar studies of other regions may be anticipated. Thus New England and the Pacific Northwest have already formed Regional Planning Commissions and comprehensive studies from the Tennessee Valley Authority may be expected.

Very properly such regional analysis must build its foundations upon studies of the physical resources and the existing economic conditions. However, before methods of improving conditions can be successfully applied, it will be necessary to understand also the cultural patterns, the ideologies, and dominant social values more or less peculiar to each region or section. Here is the field for sociological research. Obviously, any com-

[4] Turner, F. J. *The Frontier in American History.* New York: Henry Holt & Co. 1920. Pp. 158-159

[5] *Economic and Social Problems and Conditions of the Southern Appalachians.* Washington, D.C.: U. S. Department of Agriculture. Miscellaneous Publication No. 205. 1935

[6] Odum, Howard W. *Southern Regions of the United States.* Chapel Hill: University of North Carolina Press. 1936

prehensive research upon the social phenomena of a whole region will require collective effort and considerable resources. A detailed outline of such research would carry us beyond the scope of this monograph. However, it may be pointed out that the recent Bankhead-Jones Act, appropriating funds for agricultural education and research, has a clause which specifically authorizes the Secretary of Agriculture to inaugurate regional studies and to establish regional laboratories. There would seem to be no reason why this act might not be made the means of regional studies of rural social conditions as the problems involved become more clearly defined and the methods of attacking them are perfected.

In addition to such more comprehensive collective enterprises it is important to point out that many local studies may be made which will represent important contributions to such regional analyses if they are conceived and planned with that end in view. Many of the investigations suggested in this report may be made to contribute to this end and it would be superfluous to attempt to enumerate them at this point. One or two examples may suffice.

It is a matter of common observation that different types of farming have a direct effect upon patterns of family life or family relationships. A recent study of the position of the girl in the rural family[7] has broadly characterized different types of families as related to the cultural complex involved in the commercial crops upon which they depend, such as the cotton farm family, the tobacco farm family, the potato farm family, etc. Whether such generalizations are warranted, whether they are the necessary concomitants of the prevailing mode of culture of the particular crop, or to what extent they are the effect of factors in the social situation, are questions which can be

[7] Miller, Nora. *The Girl in the Rural Family.* Chapel Hill: University of North Carolina Press. 1935. Also see Vance, Rupert B. *Human Factors in Cotton Culture.* Chapel Hill: University of North Carolina Press. 1929. Chapter IX

answered satisfactorily only by several intensive studies of family life in various localities. Such studies will have to depend mainly on the case method, using a definite outline of family phenomena to be observed, although the final interpretation must needs be based on a broad interpretation of the numerical frequencies of different types of relationships. If a definite association between certain methods of production of a given crop or agricultural product and certain patterns of family life can be definitely established, it might be possible to reveal some of the factors which underlie the attitudes and values of the families engaged in these types of agriculture, and to show whether the consequences to family life are necessarily inherent in the agricultural situation or how they might be improved. Such studies would be of value as contributions to our knowledge of family life, but when a sufficient number of them had been made to validate the common conclusions, there would be a mass of evidence which would give new insight into the social characteristics and the human resources and needs of any given region.

As another example, it has been shown in the regional monographs previously cited that both in the South and in the Southern Appalachians there are relatively few farmers' organizations and few cooperative associations. Just why is this? Are the reasons purely economic, or do such factors as less education and more social stratification account for less associative effort in these than in other regions? To answer these questions satisfactorily will require various types of investigation, historical, case studies, analysis of educational statistics, etc., but as a result we might hope to gain a better understanding of what are the factors which impede the organization of farmers in these regions and what might be done to better the situation. The problem is evidently not a local one, and consequently intensive local studies which come to common conclusions would together form the basis for regional generalizations.

SOCIAL EFFECTS OF MARGINAL LAND

Prior to the industrial depression a considerable interest had been aroused in the problems of marginal and submarginal land, particularly with regard to the conservation of natural resources and reforestation, and important advances had been made in defining state policies by a few states, notably New York and Wisconsin. The problem was conceived as an economic one, the social effects of submarginal land on its inhabitants was considered only incidentally, and the federal government had assumed no responsibility in the matter outside of making purchases for national parks and forests. When the federal government undertook to aid the states in furnishing relief, it soon became apparent that there was an intimate association between the volume of relief and the nature of the land, which was demonstrated for large regions in the studies of problem areas, cited above. For the first time the nation became acutely aware that it had a problem of conserving its human resources on lands which cannot afford a reasonable standard of living, and this became increasingly apparent as the government extended its efforts in rural rehabilitation.

Outside of the regional studies mentioned in the last section, little has been done to show the exact relation between the percentage receiving relief and land utilization types in individual states, although the maps published by Kirkpatrick[8] for Wisconsin show this relation very clearly for the Cut-over section of northern Wisconsin. In Tompkins County, New York, A. B. Lewis[9] has shown that in marginal land classes I and II (the poorest classes), 11 per cent of the occupied houses had received poor relief during the 18 months January 1, 1931 to

[8] Kirkpatrick, E. L. and Boynton, Agnes M. *Wisconsin's Human and Physical Resources.* Madison: Resettlement Administration. Research Section. Region II. July 1936. Figures 14 and 36

[9] Lewis, A. B. *An Economic Study of Land Utilization in Tompkins County, New York.* Ithaca: Cornell University Agricultural Experiment Station. Bulletin 590. 1934

June 30, 1932, while in class III the percentage was 3.3 and in classes IV and V (the best classes) it was 3.8. The reconnaissance survey of land types conducted under the auspices of the National Resources Board, with the more detailed studies made in various states, furnish the basis for a more exact analysis of the relation of marginal lands to pathological living conditions in the several states. Within this broad field there are numerous problems warranting research upon some of which important beginnings have already been made.

It will be desirable to know for each section what was the origin of the population. Is it an old and established population, as in the Appalachian Highlands, which has become relatively impoverished by changes in the environment; or is it the result of recent migrations, as in the Lake States Cut-over Area or in the Short-grass Area of the Western Great Plains? In southern New York and northern Pennsylvania there has been a considerable migration to the marginal lands of foreign born persons who had previously worked in the Pennsylvania mines. This involves a study of mobility (see above) as related to land types.

What is the relation of the type of land to the sort of people living upon it? How long have they been farming? Have they changed vocations? Have they been successful or not, previously? How much schooling have they had? What is the relation of the relief rate to the type of land? Is there any relation between land type and delinquency, either of juveniles or adults? To what extent is it necessary for women and children to do work on the farm in order to make a living in relation to land types? Are there differences in size of family? An important contribution to the relation of vocational history to types of farming has been made by Hypes and Markey[10] in Connecticut, who gave

[10] Hypes, J. L. and Markey, John F. *The Genesis of Farming Occupations in Connecticut.* Storrs: Connecticut Agricultural Experiment Station. Bulletin 161. 1929

consideration to soil types but did not show definite association between soil and people. Data being collected by the Resettlement Administration with regard to their rehabilitation clients will be of value in answering these questions, but does not include all the families in a given land type area. Adequate records of the families on marginal areas being purchased by the Resettlement Administration will be of more value, but should be supplemented by similar studies of families on adjacent areas.

One of the chief objectives of the studies just outlined would be to indicate whether the people now occupying marginal lands are likely to succeed if established on better land, which is a major question in the work of the Resettlement Administration. A noteworthy attempt to answer this question for mountain whites of eastern Kentucky is being made by Messrs. Nicholls and Oyler of the Kentucky Agricultural Experiment Station in a study they are making of the success or failure of those who have migrated from mountain counties to the Kentucky lowlands. This study should throw light on this problem and those attempting research of this sort will profit by studying their methods.

Such a study at once raises the question as to whether those who migrate from marginal lands to better land may not be the better individuals. Do they represent the average on the marginal lands? Is there a selective migration of the better individuals from marginal lands to better land and into industry and other vocations? If so, does this tend toward the gradual decadence of the people living on marginal and submarginal lands?

The inadequate living derived from marginal lands is reflected in the poor schools and churches and the lack of organized social life and medical facilities in these areas. This gives rise to questions social and economic. Is it possible to rear children in such areas who, if they migrate to better areas, will not be very definitely handicapped by their lack of education and training? The exceptional ones will succeed and will be pointed out as shining examples, but what of the average individuals reared in such surroundings? Will they not tend to be less able to make satisfactory adjustments in their new environment and will not a larger proportion of them require assistance of one sort or another? Are they not likely to be more of a liability to the better areas into which they migrate than those who had a better start in life?

A study in southern Ohio or the lowlands of Kentucky might answer this question with regard to the migration into these areas of the mountaineers from eastern Kentucky and Tennessee.

If the residents on marginal lands are unable to maintain satisfactory schools, it seems necessary for the state to assist the poorer districts. But this raises an economic problem for in some cases it would be a better investment for the state to buy up the land and move the residents than to maintain the schools, even if state aid for roads and health and relief services[11] are not considered. Further studies of the cost of the necessary social services such as schools, health, and relief to the locality, the county, and the state, in typical marginal areas, are desirable to show whether it is feasible to maintain a standard of life which will not be socially dangerous.

PART TIME FARMING

A considerable proportion of the apparent increase in the number of farms during the past decade may have been due

---

[11] Lewis, A. B. *An Economic Study of Land Utilization in Tompkins County, New York*. Ithaca: Cornell University Agricultural Experiment Station. Bulletin 590. 1934. P. 33. "The state aid received in school districts wholly in land classes I and II amounted to 4.7 per cent interest on the full value of the taxable property in the districts. In about twenty years, state aid for schools in land classes I and II would amount to a sum large enough to purchase all of the taxable property in the school districts in these land classes." Young, G. E. *Marginal Farm Land in Southern Indiana*. Lafayette: Purdue University Agricultural Experiment Station. Bulletin 376. Figure 20 has shown that in some townships in southern Indiana the state aid to schools amounts to 3 per cent or over.

to city workers moving to the land and engaging in part time farming. This movement has been going on steadily since the world war and particularly in the last decade. There is no evidence that the increase was materially greater since 1930 than before then,[12] but public interest was aroused in the possibilities of part time farming as a means of economic security during a depression by the work of the Subsistence Homesteads Division and the Resettlement Administration. Davis states[13] that 63 per cent of the farms in Connecticut are operated on a part time basis. Comparisons are difficult because of differences in the definition of part time farming. The federal census of agriculture of 1935 reports only those places which were within its definition of a farm by the number of days work employed off the farm. Part time work off the farm was reported by 30 per cent of all farm operators in the United States, but nearly three-fourths of these were employed less than 150 days. Those employed off the farm for 150 days or more are definitely part time farmers, and constituted 28 per cent of all those reporting employment off the farm. "Farm operators reporting 150 or more days off the farm for pay constituted 8.4 per cent of all farm operators in the United States. More than 1 out of 5 of all farm operators in New Hampshire, Massachusetts, Rhode Island, Connecticut, and more than 1 out of 7 of all farm operators in Maine, Vermont, Virginia, West Virginia, Florida, Arizona, Washington, Oregon, and California reported 150 days or more of work for pay at jobs not connected with their farms in 1934."[14]

---

[12] Davis, I. G. and Salter, L. A. Jr. *Part Time Farming in Connecticut.* Bulletin 201. Storrs: Connecticut Agricultural Experiment Station. March 1935. Table 1. Salter, L. A., Jr. and Darling, H. D. *Part Time Farming in Connecticut.* Bulletin 204. Storrs: Connecticut Agricultural Experiment Station. July 1935. Table 5. Tate, Leland B. *The Rural Homes of City Workers and the Urban-Rural Migration.* Bulletin 595. Ithaca: Cornell University Agricultural Experiment Station. April 1934. Figure 9

[13] Davis, I. G. and Salter, L. A. *Op. cit.*

[14] "Part Time Farmers. Summary by Divisions and States." *United States Census of Agriculture, 1935.* Washington, D.C.: U. S. Department of Commerce, Bureau of Census. Press release. June 15 1936

The proportion of those employed off the farm 50 days or more increased only about 6 per cent between 1929 and 1934, while those employed 1 to 24, and 25 to 49 days increased 13 and 10 per cent respectively. These figures apply only to farm operators and do not include other members of the family who may have worked off the farm.

It is evident that part time farming is most common near cities in the manufacturing areas and in states where industries such as lumbering and mining have declined and their workers have been forced into farming. Some of the part time farmers are men who formerly devoted their full time to farming and have been forced to seek outside employment, but the great bulk of them are city or industrial workers who use part of their time and the assistance of their families in farming. In Connecticut, Davis[15] found that the previous occupation of only 12 per cent of part time farmers was farming or horticulture. This movement to the country has been motivated by desire for increased earnings and cheaper living, love of the country, search for improvement of health, and the feeling that it is a better place to rear children.[16] Economic security does not seem to have been a considerable factor in motivating this movement during the past decade or more, but the depression has called attention to the security afforded in case of unemployment and this has been a chief argument for the promotion of subsistence homesteads. Undoubtedly the experience of the depression will encourage the further growth of part time farming.

Numerous studies have shown that the income received from the farm furnishes only a fraction of the family living and is insufficient to maintain the family without income from employ-

[15] Davis, I. G. and Salter, L. A. *Op. cit.* Table 2
[16] Hood, Kenneth. *An Economic Study of Part-time Farming in the Elmira and Albany Areas of New York, 1932 and 1933.* Ithaca: Cornell University Agricultural Experiment Station. Bulletin 647. April 1936. Table 167, p. 129

ment. Thus on 725 part time farms in New York Hood[17] found that the net farm income and privileges amounted to an average of $214 as against $751 net income from work off the farm. There seems to be little evidence that part time farmers produce enough for sale to affect the market prices of farm products produced by full time farmers. Indeed they often furnish a small market for the latter.

Most of the studies of part time farming have been concerned chiefly with its economic aspects,[18] and relatively little has been done to determine the effect on social relationships.

Inasmuch as part time farming has been conceived as a means of subsistence for those unemployed, it is desirable to know whether those who have undertaken part time farming are the sort of people likely to be unemployed and receiving public relief, or are they the more enterprising and energetic workers. This question has not been answered by the reports so far published and needs further investigation in various sections where part time farming is common. In the studies of Troxell, Cottrell, Allen, and Edwards of the textile workers in the southeastern states, there were almost no part time farmers on relief and

[17] *Ibid.* Table 9, p. 19

[18] In addition to those cited above see Morison, F. L. and Sitterley, J. H. *Rural Homes for Non-Agricultural Workers.* Wooster: Ohio Agricultural Experiment Station. Bulletin 547. February 1935. Smith, F. V. and Lloyd, O. G. *Part Time Farming in Indiana.* Lafayette: Purdue University Agricultural Experiment Station. Bulletin 410. April 1936. Adams, R. L. and Wann, J. L. *Part Time Farming for Income.* California Agricultural Experiment Station. Bulletin 581. 1934. Noble, C. V. *The Cost of Living in a Small Factory Town.* Ithaca: Cornell University Agricultural Experiment Station. Bulletin 431. 1924. Rozman, D. *Part Time Farming in Massachusetts.* Massachusetts Agricultural Experiment Station. Bulletin 266. 1930. Walker, W. P. and DeVault, S. H. *Part Time Farming and Small Scale Farming in Maryland.* University of Maryland Agricultural Experiment Station. Bulletin 357, 1933. Allen, R. H., Cottrell, L. S. Jr., Troxell, W. W., Herring, Harriet L., and Edwards, A. D. *Part Time Farming in the Southeast.* Research Monograph IX. Washington, D.C.: Works Progress Administration, Division of Social Research. 1937

judged by various standards, they seemed to be rather above the average of their communities. In the two areas studied by Hood in New York he found 16 per cent of the 725 families received relief in some form, but whether this is a larger or smaller proportion than for others in the same classes of employment is not stated. Dr. N. L. Whetten writes from Connecticut:

In one or two of the industrial part-time farming communities we have evidence to show that the part-time farmers have a rather high rate of relief compared to either the full-time farmers or the non-farmers. This seems to be the opposite from what is generally believed concerning part-time farming communities. It seems that in one community particularly the subsistence from the small amount of farming was just sufficient to keep them from making drastic efforts to secure employment when their old jobs gave way. Very few of these people seemed to have gone outside of their local area to find work. They had enough to keep from starving and preferred to take a chance on getting more through governmental agencies. At the same time we know that many other workers come into this area from nearby cities and work in the local factories.

It should be noted that this is cited as exceptional, yet it establishes the point that the types of people engaged in part time farming vary considerably in different areas. In the southern Appalachian Highlands a considerable proportion of relief clients are part time farmers who get some employment in mining or lumbering. This matter of the type of people who have succeeded in part time farming is of primary importance in developing policies for providing subsistence homesteads. It would be desirable to know to what extent part time farming has enabled unemployed operators to maintain themselves without public relief or whether they required less relief than others of similar status. A considerable number of case studies in different localities would seem necessary to answer these questions.

Where part time farmers are infiltrated into an agricultural community an important question is the degree to which their families participate in local organizations and community life.

Hood[19] states that of 2,403 families studied in six counties in New York State, three-fourths of the affiliations of operators were with organizations in the country communities in which they lived, as against about one-fourth in city organizations, and this is confirmed by Tate's[20] study of Monroe County. Among the textile workers studied by Troxell, Cottrell, Allen, and Edwards the families of part time farmers participated in local organizations rather more than others and furnished a larger proportion of leadership.

The facility with which part time farmers are assimilated into the life of the local community depends upon whether they are part of the native culture or foreign to it, and to the class of people involved. Hood[21] gives an interesting account of the attitude of New York farmers to the part time farmers, which is generally favorable. In some places they are welcomed because they furnish a source of occasional labor during the peak periods when farms require extra labor. Case studies of individual communities in which there has been a considerable immigration of part time farmers would be valuable in showing their influence on community life.

One aspect of the immigration of part time farmers which warrants special investigation in community case studies is the question as to whether advantages accrue to the community from an influx of such farmers. The addition of a number of part time farmers furnishes a broader tax base and may facilitate obtaining better schools, roads, electricity, and telephone service. Their immigration furnishes a market for real estate which may tend to raise values of all property. If the newcomers are very poor, however, they may only increase the burden on the com-

---

[19] Hood, Kenneth. *An Economic Study of Part Time Farming in the Elmira and Albany Areas of New York, 1932 and 1933.* Ithaca: Cornell University Agricultural Experiment Station. Bulletin 647. April 1936. P. 136. Table 174

[20] *Op. cit.* P. 31

[21] *Op. cit.* Pp. 134-135

munity's services. Thus, whether or not such immigration is an asset or a liability depends upon the types of people and their standard of living.

There are two classes of part time farmers who may become a problem for the local community in times of depression. One consists of those families whose income and standard of living is low and who are most likely to be thrown out of employment, such as the cases described by Dr. Whetten in Connecticut. The other consists of people who are approaching the retirement age and who are buying a little place so that they can add to their income when employment is slack and will have a place to live when they are too old for steady work. If old age pension systems require local support, some suburban counties may find that they are paying the cost of retiring city workers, just as they are now paying for the education of children who become city workers. In such an event it would seem only just that the state should assume the entire cost as the only means of equalization.

Two reasons often advanced for moving from the city to a part time farm are that it is a more healthful environment and a better place to rear children. These are common opinions, but no definite evidence has been assembled to show their validity. It would be very desirable to make a number of rather intimate case studies to show whether, and under what conditions, health has been improved or health conditions are superior or inferior to those in the city, and whether family relationships are better and the opportunity for the personality development of children is better. It is perhaps too late to get at the effects of the depression in these matters but some efforts in this direction should prove worthwhile.[22] Research of this type would require a con-

[22] See Collins, Selwyn D., and Tibbitts, Clark. *Research Memorandum on Social Aspects of Health in the Depression;* Stouffer, S. A., and Lazarsfeld, Paul F. *Research Memorandum on the Family in the Depression.* (monographs in this series)

siderable number of case studies and very careful analysis and interpretation, but would be of fundamental importance in showing whether these assumed advantages actually occur and under what conditions they do or do not exist.

REHABILITATION AND RESETTLEMENT

As soon as the Federal Emergency Relief Administration was fairly well organized, it became aware that unless something were done to enable rural families to again become self-supporting many of them would be on relief indefinitely, and that some aid for rehabilitation or resettlement was necessary. As facts were gathered about the problem areas, the conditions on marginal lands, and the trends in population growth and mobility this need became more evident. It is not necessary to recount the history of the government's experiments in this field, which have now been centralized in the Resettlement Administration. Our interest is rather to indicate the research problems which have arisen in developing governmental policies and methods in this field, what research has shown concerning the factors involved, and what remains to be done for the discovery of those facts which will be of use in preventing the recurrence of a similar situation or in dealing with it more effectively if and when it occurs.

In general, rehabilitation consists of financial aid and supervision in helping rural families to become self-supporting in their own residence or in the same community; while resettlement is resorted to only for what are termed "stranded families" who live in areas where there is no probability of their being able to regain their former employment or where the land is unsuitable for successful farming. In either case, it is desirable to know what proportion and what sort of families on relief will probably profit by these types of assistance and the number and kind that are not qualified for aid. To determine these facts the Rural Research Unit of the Division of Research, Statistics, and Finance of the Federal Emergency Relief Administration (and its successor, the Division of Social Research of the Works Progress Administration) has made extensive investigations by its own staff and in cooperation with the state relief administrations, through employing the rural sociologists and economists of the state agricultural colleges and experiment stations as state supervisors of rural research. As a result a large mass of data on the characteristics of rural families which received relief has been brought together. These data have been published by FERA and WPA in a series of mimeographed bulletins and in monograph form,[23] and by state agricultural experiment stations[24] and by state relief administrations in bulletins embodying the results for their states. To attempt any analysis or abstract of this literature is outside the province of this report, but it is important to note that as a result of these investigations we now have a picture, which had never been portrayed before, of the conditions in rural life which have made relief necessary, of the kinds of families who need relief, and of the volume of relief necessary for certain classes. The number of cases in which the head of the household was ineligible for rehabilitation by reason of old age, of poor health, of being crippled or otherwise incapacitated, or of being a widow with dependent minor children,

[23] Beck, P. G. and Forster, M. C. *Six Rural Problem Areas*. Research Monograph I. Washington: Federal Emergency Relief Administration. Division of Research, Statistics, and Finance. 1935; McCormick, Thomas C. *Comparative Study of Rural Relief and Non-Relief Households*. Research Monograph II. Washington: Works Progress Administration. Division of Social Research. 1935; Woofter, T. J., Jr. *Landlord and Tenant on the Cotton Plantation*. Research Monograph V. Washington: Works Progress Administration. Division of Social Research. 1936

[24] Landis, Paul H. *Rural Relief in South Dakota*. South Dakota State College Agricultural Experiment Station. Bulletin 289. June 1934; Landis, Paul H., Pritchard, Mae, and Brooks, Melvin. *Rural Emergency Relief in Washington*. State College of Washington Agricultural Experiment Station. Bulletin 334. July 1936. A considerable list of mimeographed reports have been issued for limited distribution.

has revealed to many states and counties the inadequate aid or entire lack of assistance for such families prior to federal relief. These data were of very practical use to the Committee on Economic Security in preparing its recommendation for the Social Security Act, and have been useful for educating the public to the need for more adequate assistance for certain classes of dependents irrespective of the depression. Furthermore, it has revealed the deplorable housing conditions of the poorer farmers and their inadequate schooling as a cause of dependency throughout the South and in some sections of the North. Prior to the depression the rural counties of most of the states had very inadequate organization for the care of dependents and were quite unaware of their needs. Federal aid for relief, whatever the local reaction to it, has forced the consideration of the better care of dependent families in rural counties.

With the return of normal economic conditions it is important that these studies of rural dependent families be continued so that we may have precise knowledge as to the number and kinds of dependent families in rural areas. There are rural slums in most sections of the country which require intelligent treatment for the elimination of poverty and delinquency for the interest of the whole community, just as do slums in cities.

The studies mentioned above have revealed the classes and types which are presumably susceptible of rehabilitation, have shown the nature and volume of rehabilitation or resettlement required in individual areas,[25] and have contributed a knowledge

---

[25] For good examples of such studies see Kirkpatrick, E. L. and Ferguson, Winifred. *Survey of Relief with Reference to Rural Rehabilitation*. Madison: Wisconsin Emergency Relief Administration. September 1 1935. Pp. 67. Mimeo. Kirkpatrick, E. L. and Thomas, Ruth M. *Resettlement and Rehabilitation in the Crandon Land Purchase Area*. Madison: Resettlement Administration. Wisconsin Rural Rehabilitation Division. August 15 1935. Pp. 45. Mimeo. Kirkpatrick, E. L., Kraenzel, Carl F., and Thomas, Ruth M. *Resettlement and Rehabilitation in the Central Wisconsin Nesting Area*. Madison: Resettlement Administration. Wisconsin Rural Rehabilitation Division. October 15 1935. Pp. 37. Mimeo.

of the standards of living necessary for resettlement.[26] In the actual process of resettling individual cases a clinical study of each family is necessary to determine its capacities and what aid will be best. Such information is being accumulated by the Resettlement Administration for thousands of rural families. It is important that careful research be made as to the sort of information which is most valuable and as to the best methods for obtaining it, for the purposes of clinical diagnosis of such cases. Although the immediate assistance of these families must go on as best it may and cannot await such studies, it would be unfortunate if the opportunity to work out the best possible methods of diagnosis were neglected. The methods of social diagnosis developed by urban social workers should furnish the background for such a technique,[27] but whereas the problems of their cases have been chiefly related to unemployment or insufficient wages, the problem of the farm family is more intimately related to problems of farm management. Not only is research on the methods of diagnosis important, but it is important that adequate case records be kept so that studies may be made of the results obtained and the effectiveness of methods used with different types of families under different conditions. This is a field of work in which the investigator must have adequate training both in social diagnosis and in the techniques of farm and home management, a qualification rarely possessed by those administering such work. Many of the better workers in this field are well aware of their lack of training in methods of diagnosis and making case records and are eager for assistance, but it can be given only as a few individuals are given opportunity to cooperate with them in intensive research which may

---

[26] Kirkpatrick, E. L. *Needed Standards of Living for Rural Resettlement.* Madison: Resettlement Administration. Wisconsin Rehabilitation Division. May 1936. Pp. 62. Mimeo.

[27] See Chapin, F. Stuart and Queen, Stuart A. *Research Memorandum on Social Work in the Depression.* (monograph in this series)

reveal the best methods and which may make possible an interpretation and evaluation of the methods used.

When resettlement is undertaken, the researches already discussed concerning population trends and mobility and conditions on marginal lands are of fundamental importance and
need to be carried further as a basis for sound policies. Two methods of resettlement are being attempted: one in which the families are moved to individual farms in an area of better land,
infiltrating them among established farms; while in the other
the resettlement is made in colonies either in or around a newly
established village. We are not here concerned with attempting
to pass on the relative merits of these two policies, upon which
there exists a decided difference of opinion, but to point out
the need for research concerning what happens in both procedures.

When the infiltration policy is pursued, most of what has been
suggested as to the study of the assimilation and accommodation
of part time farmers will be equally applicable to research concerning families relocated in this manner. Where resettlement
is made in colonies the opportunity and need for sociological research is much more acute. Here are attempts to create new
communities and to establish suddenly by artificial means the
relationships which ordinarily arise in community development
by the slow processes of accommodation and integration. The
attempts to give the managers of such colonies some training
in the methods which would seem most desirable have brought
out the meager amount of knowledge which is available for
their use, and have revealed the importance of making a careful
analysis of the experience of these communities as a means for
the better guidance of such experiments in the future.

All that has been said above with regard to investigation of
the diagnosis and selection of families is peculiarly important
in the establishment of such colonies, and some research in this
field is now being carried on by the Division of Farm Population

and Rural Life of the U. S. Bureau of Agricultural Economics
by Dr. Marie Jasny.[28]

With the establishment of a colony the whole history of the
process of community organization, the various groups and fac-
tions, the growth and participation in organizations, leadership,
cooperation, and conflict, should be recorded and interpreted.
Here is an opportunity for important research by those institutions or individuals who are located near enough to such colonies
to make frequent observation possible. Although a study of
some of these colonies is now being undertaken by federal research workers, it is very questionable whether it will ever be
possible for them to publish a thoroughgoing analysis of these
projects, no matter how objective or satisfactory their research
may be. Indeed the probability is that the most important findings are not likely to be made available to the public. This is
merely to say that governmental agencies, irrespective of the
administration in power, are not likely to permit the publication of findings which reflect on their own work or that of other
departments of government.[29] For this reason it will be very
fortunate if carefully planned research concerning the social
structure and processes of these colonies can be made under
private auspices. Indeed the Committee on Social and Economic
Research in Agriculture of the Social Science Research Council
prepared and recommended such a research project in 1934, with
the approval of federal administrative officers concerned, but
funds were not then available to undertake it. The outline prepared for this project is available for any who may be interested
in undertaking such research. It is important that research be

[28] For an announcement of this and tentative conclusions see *Farm Population
and Rural Life Activities*. Washington: Bureau of Agricultural Economics. 10:3.
No. 4. October 1 1936. Mimeo.

[29] The valuable reports on the Agricultural Adjustment Administration and
the National Recovery Administration made by the Brookings Institution could
not have been made by employees of those administrations.

carried on as these experiments are started and are in progress, for otherwise the most important facts will not be recorded and it will be difficult, if not impossible, to obtain them at a later time when those involved must depend on memory for them.

European countries have had a considerable experience in resettlement projects which are being investigated by the Resettlement Administration,[30] but cultural conditions are so different that although much may be learned concerning the economic and technical aspects of colonization, it is improbable that European experience will throw much light on many of the social problems which will arise in colonies in this country. Inasmuch as a considerable amount of resettlement of rural families in "stranded areas" seems inevitable, the question of whether colonies are desirable and, if so, how they may be most successfully inaugurated and operated, is one of large importance and which may have a far reaching affect on the social organization of rural life in this country if they prove successful.

SUMMARY

Some of the more important topics requiring social research which have been suggested by our consideration of the problems of agricultural readjustment are: (1) studies of factors which account for the social differences in regions, as, for example, the effect of type of farming, particularly of types with single cash crops, upon family life and family relationships, and the number of farmers' organizations and the extent of participation in

---

[30] Kraemer, Erich. *Land Settlement Technique Abroad.* "Organization of Activities in England, Germany, and Italy, Resettlement Administration." Land Policy Circular Supplement. July, 1935. *Selection of Settlers in Agricultural Settlement of Several European Countries.* Washington, D.C.: Resettlement Administration. Land Utilization Division, Land-Use Planning Section. Land-Use Planning Publication No. 5. July 1936. Mimeo; see also Loomis, Charles P. *The Modern Settlement Movement in Germany.* Washington, D.C.: U. S. Dept. of Agriculture. Bureau of Agricultural Economics. Division of Farm Population and Rural Life. February 1935. Pp. 68. Mimeo. Includes bibliography

them; (2) investigations of the types of people who live on marginal and submarginal lands and of the cost of maintaining various public services in these areas; (3) studies of the types of people who succeed in part time farming, their effect on community life, both as regards their assimilation and the increased support of community institutions, and the effect of part time farming on the health of members of the family and their family relationships; (4) studies of case histories of families being rehabilitated as a basis for establishing methods of diagnosis of the capacity of clients for various types of rehabilitation; (5) the study of the social processes involved in newly established colonies as a mean of establishing a body of principles for the training of leaders or managers of such colonies and as a guide to future policies concerning them.

Chapter IV

Status and Stratification of Farmers

WHAT has been the influence of the depression on the social and economic status of the farmer? Has it tended to increase his importance as compared with that of other occupational classes in the public mind? Has it tended to solidify farmers as an occupational class or to increase stratification among them? These are questions which cannot be answered in general or for the country as a whole, for the situation differs by regions. Among the more important topics affecting these questions are those of tenancy, farm laborers—particularly migratory laborers—part time farming, the increase of small farms, the rise of the southern proletariat, and the possibility of a dependent class resulting from relief policies. These and other related subjects are considered in this chapter.

TENANCY[1]

There is a general impression that the depression has caused a considerable increase in the number of tenant farmers. If the gross increase is considered, this is true, for the number of tenant

---

[1] For bibliography see Bercaw, Louise O. and Hennefrund, Helen E. *Farm Tenancy in the United States, 1925-1935.* Washington, D.C.: U. S. Dept. of Agriculture. Bureau of Agricultural Economics. Agricultural Economics Bibliography No. 59. November 1935. Pp. 86. Mimeo. After this section had been written there was published the report of Woofter, T. J., Jr. *Landlord and Tenant on the Cotton Plantation.* Washington, D.C.: Works Progress Administration. Division of Social Research. Research Monograph 5. 1936. It has not been possible to digest this report for the present discussion but the student is referred to it as the most recent and thorough study of this subject.

farmers in the United States increased 7.5 per cent from 1930 to 1935. However, if the proportion of farm operators who are tenants is considered, it is found that there was an actual decrease (0.3 points) from that of 1930.[2] In both the North and the South the largest proportion of tenancy is in the Central States. In all of the North Central States the proportion of tenancy increased; but in Illinois, Wisconsin, Minnesota, and Iowa the percentage of all farms operated by tenants increased less than from 1925 to 1930. In the South Central States the percentage of tenants declined in all states except Kentucky, in which the increase was less than from 1925 to 1930, and Tennessee, which remained the same. These facts indicate that the effect of the depression on tenancy was very different in the North than in the South, and that it was different in some states than in others. The differences in the North Central States indicate the need for careful analysis of the tenancy trend in each state and by agricultural regions. It will be particularly important to isolate the influence of the increase of small farms and part time farms near cities from the trend of tenancy in the strictly agricultural counties devoted to general farming.

The importance of such an analysis has been well shown by Frey and Smith in a study of tenancy in ten southern states of 170 counties with a high proportion of croppers,[3] which gives the picture of the tenancy situation in the counties with high production of cotton. Their comparisons, however, are of the percentage of increase in the numbers of tenants of each class.

The trend of tenancy, the analysis of changes in the type and color of tenants, and a comparison with other parts of the coun-

---

[2] Tables giving the data for these statistics on tenancy are given in article by the author entitled "The Effect of the Depression on Tenancy in the Central States." *Rural Sociology*. 2:1-9. No. 1. March 1937

[3] Frey, Fred C. and Smith, T. Lynn. "The Influence of the AAA Cotton Program upon the Tenant, Cropper and Laborer." *Rural Sociology*. No. 4. December 1936

try are better made by using the percentages of all farms operated by tenants, although the contrasts would be greater if confined to the counties where there are most croppers. To reveal the factors which caused a decline of tenancy in the South from 1930 to 1935 it is necessary to separate types and color of tenants. The percentage of all farms which was operated by croppers increased from 1920 to 1930, but declined in all the South Central States except Mississippi from 1930 to 1935, whereas the percentage of other tenants increased in the last five years in all of these states except Mississippi and Louisiana. With the exception of the last two, the increase of the percentage of the other tenants was similar to that in the North Central States. If the croppers are separated by color of operator, it is found that the percentage of white croppers decreased in all states and that the percentage of colored croppers also decreased in all except Alabama and Mississippi. On the other hand, the percentage of other tenants who were white increased in all states, but declined for the colored other tenants in all states. It is evident, therefore, that the general decrease in tenancy was due to the decrease of croppers and to the decrease of colored tenants, and that the percentage of white tenants actually increased.

If the percentage of all croppers and of all other tenants is computed by color of the operator, it is found that the proportion of white croppers decreased and that of the colored croppers increased in Alabama, Mississippi, Arkansas, and Louisiana. Moreover, the proportion of other tenants who were white increased while that of those who were colored decreased in the same states. This would seem to indicate that white croppers were displaced by Negro croppers, and that Negro other tenants became croppers. Although the proportion of colored croppers increased in these four states, there must have been a considerable number of colored croppers who lost their status

as croppers, because the proportion of all tenants and of all operators (including owners) who were white increased in all states
and more decidedly in these four states.

From the above data it is evident that to determine the effect
of the depression upon changes in farm tenure will require careful analysis for each state or, preferably, for agricultural regions and particularly types of counties. To what extent these
changes were due to the depression or were the result of the
policies of the Agricultural Adjustment Administration and of
better credit facilities through the Farm Credit Administration
will probably be difficult to determine, but warrants detailed
analysis.[4]

Frey and Smith state that one effect of the AAA program has
been to decrease the mobility of cotton tenants and croppers in
the South. That there has been a decrease in mobility is shown
by the census figures of the percentage of croppers reporting the
years on the farm in 1935 as contrasted with 1930. Although
there was a slight increase in the percentage of croppers on the
farm under one year, the percentage of those who had been on
the same farm one year was much less in 1935 than in 1930,
while those on the same farm for two or more years showed a
considerable increase,[5] throughout the South. These data on the
period of farm occupancy will warrant careful study to show the
areas of low and high mobility and should make possible the
determination of just when changes of tenure within given areas
occurred.

It is a question, however, whether any statistical analysis of the

---

[4] For examples of such studies see Blackwell, Gordon W. "The Displaced
Tenant Farm Family in North Carolina." *Social Forces*. 13:65-73. October 1934;
Hoffsommer, Harold. "The AAA and the Cropper." *Social Forces*. 13:494-502.
May 1935; Johnson, Charles S., Embree, Edwin R., and Alexander, W. W. *The
Collapse of Cotton Tenancy*. Chapel Hill: University of North Carolina Press.
1935. Pp. 81

[5] U. S. Department of Commerce, Bureau of the Census. "Period of Farm
Occupancy." U. S. Census of Agriculture, 1935. Press release. October 28
1936

census figures concerning tenure will reveal what has actually happened during the depression. Thus, with the knowledge of a large number of farm foreclosures, it seems remarkable that in a state like Iowa the increase in the proportion of tenancy from 1930 to 1935 was less than in the previous five years. May it not be that a considerable number of former owners have become tenants, while former tenants have become laborers, so that in spite of the fact that their places have been taken by others rising in the scale, the relative position of the former owners, as a class, is much below that which it was prior to 1930? These facts cannot be ascertained from the published census figures, but Dr. O. E. Baker has pointed out in conversation that it should be possible to ascertain the changes in tenure of individuals by comparing the individual schedules of the agricultural censuses of 1930 and 1935. If this were done for well-selected samples of typical townships in various counties where the tenancy rate was previously high, or better where it showed a higher rate of increase from 1930 to 1935, it might be possible to show what has actually occurred in change of tenure caused by the depression.

Furthermore, as pointed out by Dr. T. J. Woofter,[6] the figures of the 1935 census of agriculture for tenants in the South must be used with great caution, and probably do not correctly depict the situation, on account of the fact that many classified as tenants were really only squatters who were living on the land rent free and making what living they could from it without having contractual relations with the landlord.

Changes of tenure are socially significant only as they indicate the probable changes in standards of living and social relations which go with them. The social aspects of tenancy in the South and in the North are radically different. It has long been held

---

[6] *Landlord and Tenant and the Cotton Plantation*. Washington, D.C.: Works Progress Administration. Division of Social Research. Research Monograph 5. 1936. Pp. 153-155

that areas with a high percentage of tenants have poorer social conditions, and there is considerable evidence to support this view. This has been clearly shown for the South by previous investigations,[7] but the effect of increasing tenancy in the better counties of the Corn Belt upon social conditions in their communities has never been accurately described or measured.

Tenancy is already an acute economic problem, but public policy with relation to it should be based upon an accurate knowledge of the social trends produced by it. For this purpose we need comprehensive studies of the social effects of tenancy in counties where it is highest in the Corn Belt as well as throughout the South. Fortunately, the Resettlement Administration has appreciated the value of such knowledge as a basis for its procedures and has undertaken a comprehensive study of the Social Correlatives of Farm Land Tenure, in cooperation with the Division of Farm Population and Rural Life of the Bureau of Agricultural Economics. Its object is "to determine how the various forms of land tenure are significantly related to the following factors: (1) family composition, characteristics, and vital processes; (2) family living, including material elements and certain types of social participation; (3) attitudes toward farm land tenure, mode of living, and other closely related matters; (4) tenure and migration histories of heads of farm households; (5) relationships between tenants and landlords, and between farm laborers and their employers; (6) social status in the community." An excellent schedule has been prepared which covers the data for the effect of tenancy on the farm family, and several thousand records are being obtained throughout the South and the Corn Belt. What is also needed is a complementary study of the effects of tenancy on the social

[7] Cf. Taylor, C. C. and Zimmerman, C. C. *Economic and Social Conditions of North Carolina Farmers*. Special Bulletin. Raleigh: North Carolina Tenancy Commission. North Carolina Board of Agriculture. 1923; Branson, E. C. and Dickey, J. A. *How Farm Tenants Live*. Chapel Hill: University of North Carolina Extension Bulletin. Vol. 2: No. 9. January 1 1923

facilities of the rural comunity, the church, the school, the library, and organizational life.[8]

FARM LABORERS, MIGRATORY LABORERS

Changes in the number and status of farm laborers are more difficult to determine because of the change of the date of the Census of Agriculture from April in 1930 to January in 1935, as a result of which the number of laborers is not comparable. Frey and Smith state their belief, based on observation and interviews with those most familiar with the situation, that the displacement of laborers by the cotton acreage reduction program has been very great and has affected them more than croppers or tenants. To the extent that southern croppers have been displaced, they have undoubtedly increased the supply of farm labor and have aggravated the labor situation. Probably only field surveys will give the necessary data for any measurement of the movements.

Formerly, there was a large amount of itinerant farm labor employed throughout the grain belt during harvest, but the advent of the combine has displaced most of it, and most farms will depend for their labor supply during the next few years on older sons and neighbors' sons who cannot find work in the city.

The depression did bring out the acute situation of farm laborers on the Pacific Coast, particularly of itinerant or migratory laborers in California, both because of the public relief necessary and the strikes among them. The considerable migration to California and the Southwest which occurred during the depression, and which caused California to restrict immigration of those without employment, is well known and has been measured by the Works Progress Administration.[9] The whole

---

[8] See other monographs in this series for general research suggestions in these and other fields.

[9] *Cf.* Webb, John N. *The Transient Unemployed.* Washington: Works Prog-

history and present status of migratory labor in California is given in a very complete report made by the State Relief Administration.[10] It is also being studied by Professor Paul S. Taylor of the University of California, who finds that it is associated with the rapid increase of intensive commercial agriculture on irrigated land and the consequent requirement of seasonal labor.[11] This situation was greatly aggravated by the immigration from drought regions, particularly from Texas, Oklahoma, and Arizona. Professor Taylor states that "during the four months ending October 15 [1935], more than 30,500 men, women and children—members of parties 'in need of manual employment' —entered California in motor vehicles bearing out-of-state licenses." But he adds that this is no problem, for "in 1927 the State Department of Education enumerated 37,000 migratory children alone." "The best present estimates place the number of men, women and children who migrate at some time during the year to work in the crops of California at from 150,000 to 200,000."[12] He draws a very vivid picture of the bad sanitary conditions and the hazards to the educational and moral development of children which occur in the camps of migrant laborers. If ways could be devised for following the careers of children of some of these itinerant families with individual case studies, they might be of great value in showing their cost to the community in the control of delinquency, and for relief and health expenditures. Examples of such studies are given in the report on migratory labor in California, by the California State Relief

ress Administration. Division of Social Research. Research Monograph 3. 1935. Map 7

[10] *Migratory Labor in California.* San Francisco: State Relief Administration of California. Div. of Special Surveys and Studies. July 1936. Pp. 224. Mimeo.

[11] *Cf.* Taylor, Paul S. and Vasey, Tom. "Historical Background of California Farm Labor." *Rural Sociology.* 1:281-295. September 1936; "Contemporary Background of California Farm Labor." *Rural Sociology.* 1:401-419. December 1936

[12] Taylor, Paul S. *Synopsis of Survey of Migratory Labor Problems in California.* Berkeley: Resettlement Administration. Undated—probably late 1935. Mimeo.

Administration (pp. 133-200). If it could be shown that such squalid living conditions cause increased taxes for the amelioration of conditions resulting from them, the consciences of the taxpayers might be aroused through their pocketbooks.

Similar problems of migratory labor occur in Oregon, Washington, and Florida. Fortunately the Resettlement Administration, in cooperation with the Bureau of Agricultural Economics, is making an extended study of farm labor problems in 11 counties in as many states, but the time is opportune for intensive studies of the condition of farm laborers in all parts of the country where intensive agriculture is dominant, and they will be of immediate value to relief administrators.

PART TIME FARMING

The social significance of part time farming has been considered above, but should be mentioned here as a factor tending toward the stratification of farmers as a class. The attitudes of part time farmers whose income is chiefly from wages earned in town or city are probably more like those of their fellow employees than of their farmer neighbors. Just what these differences are, how they affect the assimilation of part time farmers in rural communities, and to what extent they accommodate themselves to the established attitudes of the community or attempt to change them, are topics worthy of investigaton. The rôle of the part time farmer in labor unions and farmer organizations, and his influence on local political attitudes, invite exploration. In a way, part time farmers and the members of their families may be in a marginal situation in their social relations, being neither townspeople nor real farmers. This indefiniteness of social status may have real significance in its effect on the personality development of their children.

SMALL FARMS

Whether they are operated by part time farmers or by those who have no other income, the considerable increase in small farms in some states between 1930 and 1935 will have the same

general effect of bringing a new element into the farming community. This has been well brought out by the data obtained
by Lively and Foott in Ohio with regard to the shifts into agriculture within the ten townships studied.

In 1928, the number of male heads of families, exclusive of farm
operators, was 1,253. Some of these were farm laborers and others no
doubt had at some time operated a farm. During the period 1928 to
1935, 18.8 per cent of this group became farm operators, i.e., owners,
managers or tenants. This was equivalent to 21 per cent of the non-relief
group and 9 per cent of the relief group. Thus the movement into farming was more particularly from the non-relief than from the relief
group. During the same period the 1928 group of farm operators lost
7.7 per cent of their numbers to non-agricultural occupations or to the
ranks of the unemployed. The loss here was especially heavy in the relief
group where 22 per cent of the operators left farming as compared with
6 per cent of the non-relief operators. Thus, it is seen that the net movement of family heads was toward farming; for while the farm operators
of 1928 lost nearly 8 per cent of their numbers, they gained 20 per
cent from other sources. That is to say, that during the seven-year
period under consideration, every farm operator that left farming in
these sample areas was replaced by three persons who were not farm
operators in 1928. Since the heaviest acquisitions to the farm operator
group came from such occupational classes as the skilled, semi-skilled
and unskilled, classes in which economic resources are likely to be
meagre it appears that some development of small-scale proprietary
farming is indicated.[13]

The last suggestion is borne out by the 1935 Census of Agriculture, which shows that while the number of farms in Ohio
increased 16 per cent from 1930 to 1935, 80 per cent of this
increase was of farms under 50 acres. Furthermore, these small
farms were mostly not those of part time farmers, although with
better times they may seek outside employment, for the number
of farms under 50 acres increased 50 per cent, while the number
of all farmers reporting days worked off the farm increased only
18 per cent. It will, therefore, be desirable to make a detailed
analysis of the increase of small farms in those states or sections

[13] Lively, C. E. and Foott, Frances. *Population Mobility in Selected Areas of
Rural Ohio, 1928-1935*. Wooster: Ohio Agricultural Experiment Station. Bulletin
582. June 1937

where this has been noticeable. In any contemplated studies, however, the possibility of better enumeration of small farms in 1935 than 1930, must not be overlooked.[14]

THE SOUTHERN PROLETARIAT

The depression has probably tended to increase the economic stratification which already existed throughout the South and seems to have further separated tenants and laborers from the landlords. This antagonism has been played upon and strengthened by the appeals of a certain type of clever politician from Ben Tillman to Huey Long.[15] There is considerable evidence that the cotton acreage reduction program of the AAA benefited the entrepreneurs much more than the tenants, and undoubtedly caused the displacement of a certain number of croppers. As a result there has been the rise of the Southern Tenant Farmers' Union,[16] including both white and Negro croppers, with croppers' strikes and serious unrest. The fact that white and Negro croppers will associate in such an organization indicates the strength of their common antagonism to the landlords, for heretofore the relationship between them was competitive.

What the future influence of this rising proletariat may be cannot be prophesied, but there is considerable evidence that the depression has helped to integrate it.[17] Local studies of the

[14] See Thompson, Warren S. *Research Memorandum on Internal Migration in the Depression.* Chapter III

[15] Basso, H. "Huey Long and His Background." *Harper's Magazine.* 170:663-673. May 1935; Simkin, Francis B. *The Tillman Movement in South Carolina.* Durham, N.C.: Duke University Press. 1926

[16] First organized in Arkansas in the late summer of 1934 according to Mitchell, H. L. "Organizing Southern Share Croppers." *New Republic.* 80:217. October 3 1934. In October 1936 the union claimed 20,000 members in 215 locals in Arkansas and 8,400 members in 74 locals in Oklahoma, besides membership in Mississippi, Tennessee, Missouri, and Texas.

[17] Thus one southern correspondent who is a keen observer of social relations writes: "The awakening of the common man, especially the 'poor white,' in southern areas is a recent development which has tremendous implications. Unless this factor is taken into account, it is impossible to gain any real comprehension of a political situation in the South. This awakening is greatly tied up, of course,

voting of white tenants and croppers and the voting of their representatives in state legislatures might furnish interesting evidence with regard to their solidarity.[18] The main point is that during the depression the economic strata of southern farmers have been further separated, and with the declining market for cotton the landlord class may not be able to maintain its traditional position. Southern landlords and politicians naturally resent the investigation of this problem by outsiders, but if some of their more intelligent leaders would become convinced that they must face the facts for their own economic interests and would give encouragement to a thorough investigation by southern economists and sociologists, it might result in defining the situation in a new light which might give rise to intelligent discussion and well considered action. A study of the status of the southern white tenant similar to that made by Charles S. Johnson of the Negro cropper in *The Shadow of the Plantation* would be most opportune.

A DEPENDENT CLASS RESULTING FROM RELIEF POLICIES

There is definite and credible evidence that in several sections dependency has been encouraged and a dependent class has arisen as the result of the methods of distributing relief. This has not been reported to any extent in the areas requiring relief on account of drought, but has occurred mostly in sections, as in the Southern Appalachian Highlands and the Lake-States Cut-over region, whose economic resources had been previously exploited so that many of the people had but a marginal standard of living.

with all such institutional changes as increased public education. The increase of communication and transportation facilities has also played an important part. It is my conviction that the developments along this line are in their beginning stages."

[18] For suggestions as to methods of analysis see Rice, Stuart A. *Farmers and Workers in American Politics.* New York: Columbia University. Studies in History, Economics, and Public Law. No. 253. 1924

At least this opinion is advanced by various individuals who are familiar with local situations.

We are not concerned with the wisdom of the government's policies, or with trying to fix responsibility for their poor administration, but with whether such opinions as the above are true to the facts. It will be important to locate all counties where such conditions are suspected and to have a careful study made to ascertain how general this attitude of dependency is, among what types of families it is most prevalent, and what have been the influences fostering it. For this purpose, it would be desirable to have careful case studies made of a considerable number of families to show the process by which their attitudes toward relief have changed and the effect on their morale and future prospects. The self-reliance and self-respect of a community, as of an individual, may be quickly lost, but will require years to regain. The sooner the situation is correctly diagnosed and measures for its correction are applied, the better the chances of recovery.

GENERAL CONSIDERATIONS

Our discussion has revealed that the question of the effect of the depression on the status of the farmer cannot be answered for the country as a whole because of marked regional differentials. However, when we attempt to estimate the status of the farmer in public opinion, there are many indications of distinct gain and of a larger solidarity among farmers. Thus, both political parties and the business world seem agreed that some measure of governmental assistance, to give the farmer an income above that which he can obtain from the prices of the world market for his products, is necessary and justified in order to maintain a satisfactory standard of living and to give him a fair share of the national income. That this opinion was non-existent ten years ago appears to be the case from the bitter opposition to the McNary-Haugen bill and its defeat by presidential veto.

The unprecedented droughts have forced an appreciation of the risks in agriculture which are beyond the farmer's control, and if a financially sound plan of crop insurance could be worked out it seems likely that it would gradually win general support. The national farmers' organizations have worked together in support of federal legislation for agriculture as never before, and the established organizations have gained public confidence and prestige in spite of—or possibly because of a contrast with— the efforts of evanescent direct action movements such as the Farmers' Holiday Association in the Corn Belt. A study of the change in public opinion toward the economic and social problems of agriculture through a study of editorial opinion in the daily press and in financial and trade journals and magazines, along the lines of that made by Dr. Hornell Hart for the Committee on Social Trends,[19] would certainly be of historical value, and might reveal what factors or methods have been most influential in changing a former antagonism to a better understanding and a greater willingness to deal with them justly.

Although there is evidence that the status of agriculture as an industry has been raised in public opinion, and that—as shown in Chapter VIII—there is more class solidarity among the better farmers, it is a question as to whether this is true of all farmers or whether a more definite cleavage has not been developed between the better class of farmers and those who are getting a bare existence from the land.

Our brief survey of what has happened to the status and stratification of farmers during the depression has shown that important changes have occurred, but that they are different for

---

[19] Hart, Hornell. *Recent Social Trends in the United States.* New York: McGraw-Hill Book Co. 1933. Vol. I. Chapter VIII. "Changing Social Attitudes and Interests."

different regions and different types of farmers, and that investigation of many phases of this topic are necessary in order to determine the changes accurately and to evaluate them. The more important subjects for research include: (1) a study of changes in the amount of tenancy by counties so as to relate tenancy to agricultural regions and crop areas; (2) the effect of high proportions of tenants on the social institutions of rural communities in the Corn Belt; (3) case studies of the families of migratory laborers; (4) the social status of the families of part time farmers in rural communities; (5) the increase of small farms by various types of areas; (6) the social status of white tenants in the South and their political behavior; (7) case studies of families in areas where there seems to be a tendency for their independence to be lowered by the general acceptance of public relief.

Chapter V

New Problems of Rural Youth

THE discussion of population problems has already shown that more farm youth from 20 to 29 years of age migrated to cities from 1920 to 1930 than of any other ten year age class.[1] During the depression this migration was checked and, as a result, there was a large increase in the proportion of youth in rural communities who, with no certain prospect of employment, face many problems of adjustment. This situation has focused attention, as never before, on the responsibility of the rural community for the needs of its youth. The problem is the more serious because the increase of rural youth, both as a result of lessened migration and because of higher birth rates, is largest in the poorer sections which are least able to furnish them opportunities, and because many of the better communities have not become aware of the fact that they face a new situation.

A preliminary analysis of the research needed in this field has been presented by Dr. Bruce L. Melvin[2] and several studies have already been made concerning the needs of rural youth.[3] The

---

[1] *Cf.* Dorn, Harold F. and Lorimer, Frank. "Migration, Reproduction, and Population Adjustment." *Annals of the American Academy of Political and Social Science.* 188:280-289. November 1936. Table II

[2] Melvin, Bruce L. "Scope of Research on Rural Youth Needed Today." *Social Forces.* 15:55-58. October 1936

[3] Thurow, Mildred B. *Interests, Activities, and Problems of Rural Young Folk. I. Women 15 to 29 Years of Age.* Ithaca: Cornell University Agricultural Experiment Station. Bulletin 617. December 1934

Anderson, W. A. and Kerns, Willis. *Interests, Activities, and Problems of Rural Young Folk. II. Men 15 to 29 Years of Age.* Ithaca: Cornell University Agricultural Experiment Station. Bulletin 631. May 1935

Anderson, W. A. *Rural Youth: Activities, Interests, and Problems. I. Married*

American Youth Commission of the American Council on Education is undertaking an extensive study of youth problems including those of rural young people.

The basic question is how many young people have remained in rural communities during the industrial depression who in normal times would have migrated, and in what areas are they most numerous. Unfortunately, the 1935 Census of Agriculture omitted practically all the questions on farm population as originally planned, so that it gives us no help with regard to this

Young Men and Women, 15 to 29 Years of Age. Ithaca: Cornell University Agricultural Experiment Station. Bulletin 649. May 1936; *II. Unmarried Young Men and Women, 15 to 29 Years of Age.* Ithaca: Cornell University Agricultural Experiment Station. Bulletin 661. January 1937

Frayser, Mary E. *Attitudes of High School Seniors toward Farming and Other Vocations.* Clemson College: South Carolina Agricultural Experiment Station. Bulletin 302. 1935. Pp. 32

Kirkpatrick, E. L. and Boynton, Agnes. *Interests and Needs of the Rural Youth in Wood County, Wisconsin.* Madison: University of Wisconsin Agricultural Extension Service. Special Circular. January 1936. Pp. 12. Mimeo

Wileden, A. F. *What Douglas County Young People Want and What They Are Doing about It.* Madison: University of Wisconsin Agricultural Extension Service. Special Circular. December 1935. Pp. 12. Mimeo

Hypes, J. L., Rapport, V. A., and Kennedy, E. M. *Connecticut Rural Youth and Farming Occupations.* Storrs: Connecticut Agricultural Experiment Station. Bulletin 182. November 1932

Lively, C. E. and Miller, L. J. *Rural Young People 16 to 24 Years of Age.* Columbus: Ohio State University and Ohio Agricultural Experiment Station. Department of Rural Economics. Bulletin No. 73. July 1934. Mimeo.

For an excellent bibliography of the whole subject see Colvin, Esther M. *Farm Youth in the United States.* Washington, D.C.: U. S. Department of Agriculture. Bureau of Agricultural Economics. Agricultural Economics Bibliography No. 65. June 1936. Pp. 200. Mimeo.

Brundage, A. J. and Wilson, M. C. *Situations, Problems, and Interests of Unmarried Rural Young People 16-25 Years of Age: Survey of Five Connecticut Townships, 1934.* Washington, D.C.: U. S. Department of Agriculture. Extension Service Circular 239. April 1936. Pp. 47. Mimeo.

For predepression studies see Kirkpatrick, E. L. *Attitudes and Problems of Farm Youth.* Washington, D.C.: U. S. Department of Agriculture. Extension Service Circular 46. 1927. Mimeo; American Country Life Association. *Farm Youth.* Proceedings 9th National Country Life Conference. Chicago: University of Chicago Press. 1927

problem, which is certainly as important as how many live stock remain on the farms. School censuses might be used to give an index of the increase of those in the older school ages, 15 to 21, but the chief migration has been by those 20 to 29 years of age. One of the best sources of reliable information will be the mobility studies previously mentioned,[4] from which it will be possible to determine the decrease in migration during the last five years. Another source of information are the various local population censuses taken at various periods since 1930.[5] In ten townships in Ohio, Lively and Foott[6] found an increase between 1930 and 1935 of 10.4 per cent in youths of 15 to 24 years of age and 6.4 per cent for ages 25 to 34. To bring the situation to the attention of individual communities it might be valuable to obtain the cooperation of high school principals in a number of rural communities, representing different types of agriculture and economic status, in order to make a census of the graduates or those who had dropped out of high school during the last five years. In the high schools of 140 villages resurveyed in 1936 Dr. E. deS. Brunner[7] finds that, of the 1935 graduates, 25 per cent were employed, 35 per cent had continued their education, and 40 per cent were unemployed. The proportion unemployed was the highest in the midwestern states, where it was 51 per cent, and where only 25 per cent had continued their education. The residence of all graduates from 30 of these rural high schools from 1930 to 1935 was obtained by Dr. B. L. Melvin, who reports that 75 per cent of them were living in rural terri-

---

[4] *Cf.* Hamilton, C. Horace. "The Annual Rate of Departure of Rural Youths from their Parental Homes." *Rural Sociology.* 1:164-179. June 1936

[5] See appendix B in *Research Memorandum on the Family in the Depression* by Stouffer, Samuel A. and Lazarsfeld, Paul F.

[6] Lively and Foott. *Population Mobility in Selected Areas of Rural Ohio, 1928-1935.* Wooster: Ohio Agricultural Experiment Station. Bulletin 582. June 1937

[7] Brunner, E. deS. and Lorge, Irving. *Rural Trends During Depression Years, 1930-1936.* Chapter VII. New York: Columbia University Press. 1937

tory in 1936, or roughly twice as many as would have remained in rural territory prior to 1930.

In addition to determining the number employed, unemployed, and in school, such a census should ascertain what vocational training the young people have had and whether they have any vocational plans, for the studies made indicate that lack of vocational training and guidance are among their chief handicaps. Thus Anderson found that, of the married young people, 82 per cent of the men and 67 per cent of the women had received no vocational training of any sort,[8] while in another county more than one-half of the young men stated that they had no vocational plan for the next five years and 48 per cent said that they had chosen no lifework.[9]

Out of this situation arises the question of how rural youth may be given vocational training. The Smith-Hughes Act furnishes federal aid for courses in agriculture and home economics which are the vocational subjects taught most commonly in rural high schools, and commercial courses are being introduced in many larger villages. For other types of vocational training there are not a sufficient number of students in the rural high school to warrant such specialization. How to furnish the rural boy and girl the variety of vocational training offered by the technical high school in the city is, as yet, an unsolved problem which is challenging educators. The need for vocational guidance has also been made more evident, and efforts to meet this need are being made by the more progressive rural high schools. There will also be increasing opportunity for those who are qualified to furnish guidance to rural youth in personal problems, for lack of vocational opportunity often creates a sense of frustration which may have a serious effect on personality at the critical age.

[8] Anderson, W. A. *Op. cit.* Bulletin 649. P. 15
[9] Anderson, W. A. and Kerns, Willis. *Op. cit.* Bulletin 631. P. 41

The CCC camps have furnished a suggestion as to a method whereby some of these problems might be met, for many observers of their work have raised the question whether some permanent institution of this sort, which was organized specifically for educational rather than for relief purposes, might not be the means of giving vocational and character training to certain types of young men who are not particularly interested in the ordinary high school curriculum. These are questions whose investigation is within the field of the science of education and will require its techniques.[10]

Both vocational training and vocational guidance[11] will need to be based upon a knowledge of vocational opportunities, so that extensive studies of vocational opportunities both in rural communities and in cities, and of the qualifications necessary for these occupations, are necessary as never before.

The basic problem of rural youth is to find employment, and pertinent research is chiefly within the field of economics. It is, of course, possible for the farm youth to work on the home farm, but if he does so in most cases either he will work for his keep or the wages will be too small to enable him to get ahead and become independent or to permit him to marry. It would seem that, if the stoppage of migration continues, there must be a definite system of wages or of partnership on the home farm, or a division of the farm so as to permit independent operation.[12]

---

[10] See also The Educational Policies Commission. *Research Memorandum on Education in the Depression.* (monograph in this series)

[11] An interesting account of how one mountain county in eastern Kentucky has tackled this problem is given by Gooch, Wilbur I. and Keller, Franklin J. "Breathitt County in the Southern Appalachians" in *Occupations: The Vocational Guidance Magazine.* 14:1011-1110. No. 9. June 1936

[12] Concerning the income of farm boys and girls, see Beers, Howard W. *The Money Income of Farm Boys in a Southern New York Dairy Region.* Ithaca: Cornell University Agricultural Experiment Station. Bulletin 512. September 1930; and *Income, Savings and Work of Boys and Girls on Farms in New York, 1930.* Bulletin 560. May 1933. For an excellent analysis of how young men become farmers, see Hypes, J. L. and Markey, John F. *The Genesis to Farming Occupa-*

In either case a lower standard of living for farm people would seem inevitable unless new markets or uses for agricultural products can be created, for with the larger use of machinery it is now possible to produce more than the market demands with the existing supply of farm labor.

It is possible, however, that in the South where there is land partly or poorly farmed which might be improved and where there is land out of production, there may be opportunity for many of these young people to enter farming. They would find it necessary to develop a type of farm on which they would obtain a large share of their living and raise sufficient cash crops to supply a modest standard of living, one which might not be inferior to what they would probably obtain as wage earners elsewhere. The possibilities in this direction may be aided by land use surveys.

These are chiefly problems of farm management and agricultural economics. The dean of one of the leading agricultural colleges of the Corn Belt has well stated the problem as the "determination of what adjustments and ideals for rural living and plans for land use management will make a place for the young people who are in excess of the needs under present conditions, but who will wish to remain in the rural communities." Dr. O. E. Baker sees the village as the hope of the future. "The solution of the problem," he says, "seems to lie in further decentralization of industry and in continued suburban development, with consequent rapid growth in the village population of the nation." . . . "The village, at least the rural village, provides for the children and the aged. If it can provide remunerative work for those in the vigor of life, the problem of maintaining a stationary, or even slowly increasing population may be solved. Moreover if factories and business establishments in the villages afforded work for the young people not needed for

tions in Connecticut. Storrs: Connecticut Agricultural Experiment Station. Bulletin 161. October 1929

farm work, much of the movement of rural wealth to the cities resulting from migration would cease. Not only the wealth now required to raise and educate children who go to the city would be retained in the rural districts, and also that transferred in the settlement of estates, but, in addition, much of the wealth accumulated by the young men and women during their lifetimes would be retained."[13] Present trends in industry[14] do not seem to give much encouragement to such a program, but it would seem worthwhile to have a more thorough study than has yet been made of the nature and circumstances of small industries which have succeeded in villages, and of the possibility of operating other industries in small units of production, but with a central marketing organization. Indeed, it might not be amiss for government to subsidize some experiments in this field for a limited time and with proper control of wages and conditions, if responsible manufacturers could be interested.

The transfer of wealth to cities through the settlement of rural estates, referred to by Dr. Baker, is a topic which needs much more thorough and extensive study than it has yet received, for the burden of paying off the other heirs is one of the major handicaps of many a young farmer.[15] A careful study of the Swiss system of settling farm estates should be made and given publicity in this country, as a means of arousing interest in this problem, solution of which will doubtless meet many legal difficulties.

One of the striking features about the migration of rural youth

[13] Baker, O. E. "Rural-Urban Migration and the National Welfare." *Annals of the Association of American Geographers.* 23:59-126. June 1933

[14] Cf. Cramer, Daniel B. "Is Industry Decentralizing?" *Study of Population Redistribution.* Bulletin III. Philadelphia: University of Pennsylvania Press. 1935; Summarized in Goodrich, Carter et al. *Migration and Economic Opportunity.* Philadelphia: University of Pennsylvania Press. 1936. Pp. 314

[15] For one of the few studies on this topic see Tetreau, E. D. *Migration of Agricultural Wealth by Inheritance—Two Ohio Counties.* Columbus: Ohio State University and Ohio Agricultural Experiment Station. Department of Rural Economics. Bulletin No. 65. September 1933. Pp. 15. Mimeo.

to the cities is that more young women than young men leave the farms. Obviously, there is less opportunity for the young women on farms, but if there is to be less employment in towns and cities the interests and needs of young women in rural life will require reconsideration. One of the important matters to be studied in this connection is the place of the young woman in the farm family and how it is affected by particular types of farming. An excellent suggestion of the scope of such a study is contained in Nora Miller's *The Girl in the Rural Family*.[16] Valid generalizations could be obtained only from a considerable number of case studies, and it would be desirable to have similar studies made in several rather similar sections within a given region and of families of different economic status.

It will also be important to make studies of the difficulties which rural youth who migrate to the cities have in adapting themselves to the urban environment. A study of the problems which they meet might be of considerable value for vocational guidance and might even affect the nature of school instruction. A beginning in this field has been made by Dr. O. Latham Hatcher[17] with regard to rural girls.

With the increased number of young people in rural communities, one of the needs noticeable everywhere is that for better facilities in recreation and leisure time activities.[18] While emigration was taking the youth from rural communities, in many cases so few were left of those beyond school age and unmarried that they had little opportunity for organized group life. The existing organizations met the needs of the two largest classes, the chil-

---

[16] Miller, Nora. *The Girl in the Rural Family.* Chapel Hill: University of North Carolina Press. 1935

[17] Hatcher, O. Latham. *Rural Girls in the City for Work.* Richmond, Va.: Garrett and Massie. 1930. *Cf.* Thompson, Warren S. *Research Memorandum on Internal Migration in the Depression.* Chapters II and IV.

[18] See Steiner, Jesse F. *Research Memorandum on Recreation in the Depression.* (monograph in this series)

dren and the older people, but the programs and activities of these organizations were not attractive to youth of 16 to 29 years of age, which, in turn, encouraged further migration. It is fortunate that this need has been recognized by the agricultural extension services of the land-grant colleges and the U. S. Department of Agriculture, and experiments are being made in many states with the organization of older rural youth. It will be desirable that these experiments be carefully followed with investigations of the efficiency of different types of programs and methods of organization, so as to reveal which are best meeting the real needs of rural youth, as contrasted with those which will seek to impose on them a program based on a priori reasoning, or which will fit into the institutional moulds of established organizations.

In this connection there is a large opportunity for a creative or experimental type of research directed toward a practicable program, rather than toward scientific generalizations, which can be carried on by the young people themselves with competent and understanding guidance, as has been illustrated by a study made of the youth of Douglas County, Wisconsin, of their own problems.[19] In such studies of the needs and attitudes of rural youth, use should be made of school teachers, who are naturally interested in these problems and many of whom will be willing to cooperate in them if rightly approached.[20] Out of many such studies and a gradual synthesis of their findings and experiences may come new ideas which would not occur to adults with a mind-set resulting from their experience in a different cultural situation.

---

[19] *Cf.* Wileden, A. F. *What Douglas County Young People Want and What They Are Doing about It.* Madison: University of Wisconsin Agricultural Extension Service. Special Circular. December 1935. Pp. 12. Mimeo

[20] *Cf.* Kirkpatrick, E. L. *What Farm Young People Like and Want.* Madison: University of Wisconsin Agricultural Extension Service. Special Circular. March 1935. Pp. 6. Mimeo

The studies made of the leisure time activities of rural youth show that reading occupies the most time, but that facilities for obtaining good reading matter are often very limited. The need for better rural library facilities, not only as a means of enjoyment but also as a means of self-education, is very apparent and their improvement is being fostered by a vigorous campaign of the American Library Association and the library departments of state governments. There is need for a detailed study of the distribution of rural libraries, their facilities, and what is needed to increase their use by rural people, in every state.[21]

As yet there seems to be no evidence of any widespread decadence in the spirit of rural youth or of a general attitude of frustration on their part, but in the studies of rural village communities made by Dr. E. deS. Brunner, it was frequently reported by the field workers that there was a noticeable increase in the number of "loafers" who had recently left school. If this tendency continues, it would seem desirable to make studies of how these unemployed young people occupy themselves. Such studies should be repeated at definite intervals to discover what changes are occurring, for it is quite possible that in such unemployed groups delinquency may develop and rural gangs may be born.

For two generations rural people have been deploring the migration of their youth to the cities, not because they saw any future for them at home, but because they disliked to sever the natural ties of affection and had a vague fear of city life. Now that rural youth do not have so much opportunity to leave the home community, there will for a time be a certain satisfaction to the older generation in keeping the young people with them, and they may fail to realize how keenly thwarted ambition is being felt and what effect this frustration is having

---

[21] See Waples, Douglas. *Research Memorandum on Social Aspects of Reading in the Depression.* (monograph in this series)

upon the formation of personality at a period when it normally develops most rapidly. Furthermore, those young people who have been compelled to return to the home community because of unemployment will have peculiar problems of adjusting themselves to the old environment. It is of the utmost importance, therefore, that local studies be inaugurated in which the people themselves participate, so that they may be made aware of the problems which their young people are facing and may do their best to give them all the help possible. Such local studies may well enlist the cooperation of all interested agencies, the school, the church, the Grange, the farm bureau, the parent-teacher associations, and kindred groups, but will need to be stimulated and directed by outside agencies such as the extension services of the land grant colleges and state universities.[22]

[22] Cf. Youth—How Communities Can Help. U. S. Office of Education. Bulletin No. 18-1. 1936

Chapter VI

Effects of the Depression on Rural Institutions

THE standards of rural life and the satisfactions peculiar to it are largely the product of its basic institutions, the family, the school, the church, and the rural community. The problems concerning them that have arisen or have been accentuated during the depression which require research to show their possible effect on the future of rural life is the subject of this chapter.

ON THE FAMILY

Whether the depression has had any effect on the rural family as an institution is difficult to ascertain.[1] The effect on the standard of living is discussed in Chapter VIII, and such changes as occurred are probably not permanent. There was a noticeable decline in the number of marriages and births in 1932 and 1933 at the depth of the depression, which has been compensated for to some extent by increases in 1935 and 1936,[2] but since this aspect of family life will be treated in another monograph of this series it need not be discussed here.[3] The noticeable decrease

---

[1] See Stouffer, Samuel A. and Lazarsfeld, Paul F. *Research Memorandum on the Family in the Depression.*

[2] *Cf.* Stouffer, Samuel A. and Spencer, Lyle M. "Marriage and Divorce in Recent Years." *The Annals of the American Academy of Political and Social Science.* 188:56-69. November 1936. A study of marriage rates through 1934 which shows a method of research on this subject is that of Hamilton, C. Horace. "The Trend of the Marriage Rate in Rural North Carolina." *Rural Sociology.* 1:452-471. December 1936

[3] See footnote 1.

in the rural birth rate, particularly in the South, would seem to indicate an increase of voluntary birth control. It is quite possible that there has been a considerable increase in the use of contraceptives, for in the last few years mail order houses have increased their sale, if one may judge by the larger space devoted to them in their catalogues. The increased mobility of rural families, particularly of the poorer class, would seem to be inimical to family life, but in many cases they were accustomed to mobile existences.

One fact brought out by the studies of families on relief is the deplorable housing conditions in many sections, but particularly in the South. The U. S. Census of 1930[4] showed that one-third of all the farm houses in the United States were valued at less than $500, and that approximately 60 per cent of the tenants in the southern states lived in houses valued at less than $500. If this was the average, what must the poorest quartile have been? In the study of rural problem areas made by the Federal Emergency Relief Administration, it was found that from 52 to 70 per cent of the families receiving relief in the 30 counties surveyed in the southern states were living in houses which were classified as unfit for human habitation on the basis of the minimum standards of the locality concerned. The percentages of relief families living in such uninhabitable houses, with the range in the counties studied, is given by areas in Table II.

These figures are bad enough, but a real appreciation of the inadequacy of the houses of southern tenants may be had by looking at pictures of typical examples such as have been presented by Rupert B. Vance.[5] In comment he says:

Some light can be shed on the level of farm housing by a consideration of the value of farm dwellings as returned by the 1930 census.

[4] *Fifteenth Census of the United States. Agriculture.* Washington: U. S. Department of Commerce, Bureau of the Census. 1930. Vol. II, Table 3

[5] *How the Other Half Is Housed: A Pictorial Record of Sub-Minimum Farm Housing in the South.* Chapel Hill: University of North Carolina Press. 1936. Southern Policy Papers No. 4

For the cotton states the average value of farm dwellings in 1930 ranged from $377 for Mississippi to $708 for Texas. Five states, Mississippi, Arkansas, Alabama, Louisiana and Georgia fall under an average of $500 per dwelling. Only Texas rises above $675. This figure indicates two things: (1) the low values of tenant cabins, and (2) the large share they contribute to the total amount of farm houses. The separation of farm owners and tenant houses in four Cotton Belt counties give an average value per house of $975 for owners and $352 for renters.

TABLE II
NUMBER AND PERCENTAGE OF RURAL RELIEF HOUSEHOLDS LIVING
IN SUBSTANDARD HOUSES: RURAL PROBLEM AREAS, JUNE, 1934[a]

AREA	NUMBER OF COUNTIES SURVEYED	RELIEF CASES LIVING IN SUBSTANDARD HOUSES		
		NUMBER	MEDIAN PERCENTAGE IN ALL COUNTIES	RANGE IN PERCENTAGES IN ALL COUNTIES
Total	42[b]	12,540	46	5–96
Appalachian-Ozark . . .	12	5,578	52	31–80
Eastern Cotton	12	3,496	60	26–95
Western Cotton	6	2,474	70	25–96
Spring Wheat	7	584	15	5–40
Winter Wheat	5	408	17	0–33

[a] Beck, P. G. and Forster, M. C. *Six Rural Problem Areas*. Research Monograph I. Washington, D.C.: FERA Division of Research, Statistics, and Finance. 1935

[b] Estimates were not available for 23 of the 65 counties in the whole survey.

Another vivid picture of the cabins of Negro tenants has been drawn by Charles S. Johnson.[6] He concludes his volume as follows:

Such dwellings and surroundings constitute a dreary setting for families and their children. One reaction to this is the constant wandering-about in search of something better, with respect to both housing and labor terms. The crowding-together of families in these small rooms destroys all privacy, dulls the desire for neatness and cleanliness, and renders virtually impossible the development of any sense of beauty. It is a cheerless condition of life with but few avenues of escape even for those who keep alive a flickering desire for something better.

Certainly, no very satisfactory family life can occur under such conditions, which are the direct result of housing standards

[6] *The Shadow of the Plantation*. Chicago: University of Chicago Press. 1934. Pp. 90-100

of the days of slavery, but which are now forced on white as well as Negro tenants. Although the Resettlement Administration is making an earnest effort to show what can be done in constructing cheap but serviceable houses, it will have but little effect on the mass of tenant housing and, as yet, no comprehensive program has been developed for alleviating this condition. Comprehensive surveys of several counties in each of the southern states to show the extent and nature of tenant housing conditions might help arouse state and federal governments to giving practical aid for their betterment.

The relief studies also brought out the need for old age pensions and mothers' pensions for widows with minor children for rural, as well as urban, families. In the rural problem areas 7 per cent of the families receiving relief consisted of broken families of a woman with children under 16 years of age,[7] and "each ten families receiving relief included an average of two persons 65 years of age and older."[8] The facts thus brought out were of assistance in showing the need for the federal Social Security Act which with the state laws accepting its provisions, will do much to help these types of families.

Whether the experiences of the depression have strengthened or weakened family ties and relationships, who can say? For those families who have been able to get along, the stresses and strains of the depression have often strengthened family solidarity, but for those who have been unable to maintain their independence they have often caused increased tensions or even family disorganization. The difficulty of obtaining any general picture of these effects is that to obtain accurate information is exceedingly expensive, for it cannot be obtained by questionnaires or short interviews, but must come through investigators

---

[7] Beck and Forster. *Six Rural Problem Areas.* Research Monograph I. Washington: Federal Emergency Relief Administration. Division of Research, Statistics, and Finance. 1935. Table 2, p. 40

[8] *Ibid.* P. 44

who have entrée to the families and who can observe their behavior for some time. A considerable amount of such data has been accumulated by competent social workers concerning rural families on relief and may be worthy of analysis where it can be obtained, but there is almost no such data for non-relief families.[9]

One change in family life caused by the depression is the larger number of children remaining at home who in normal times would have obtained work elsewhere. As it seems probable that this situation will continue irrespective of economic conditions, it would be desirable to ascertain whether the larger number of older children in the home has made for better family life or has created more tensions. If adequate data on this point could be obtained from a considerable number of families of different types, and again at five year intervals, it might throw light on the problems of older rural youth, which promise to challenge rural society for some years to come. High school principals and teachers and 4-H club agents who have known these young people and who are acquainted with their families, might be means of making contacts with them or might be employed for gathering the data.

ON THE SCHOOL

The effect of the depression on our public school system is discussed in another monograph of this series,[10] so that only those aspects of the situation which seem to be peculiarly significant for rural life will be considered.

One of the chief effects of the depression has been to bring the problems of rural schools into public attention because of the seriousness of their plight. In many cases the school terms were seriously shortened, schools were closed, and teachers' sal-

---

[9] See note 25, p. 47

[10] See Educational Policies Commission. *Research Memorandum on Education in the Depression.*

aries were reduced to below the price for ordinary labor. Conditions would have been much more serious had it not been for temporary federal aid through the FERA. The city schools also suffered, but rarely were they closed or the terms shortened. The open country schools suffered much more than the larger village schools. Thus in the 140 villages surveyed by Dr. Brunner[11] in 1936, he found that teaching costs per pupil declined 20 per cent in the villages and 30 per cent in the open country schools, and that teachers' salaries were reduced 17 per cent in the villages and 24 per cent in the open country.

The difficulties of the rural schools have been an object of concern to educational administrators for many years,[12] but the acute situation forced the attention of the general public upon them as never before. The studies of land utilization in New York State showed the high cost of maintaining one-room schools in areas of marginal land, and the unequal tax burden upon these districts which are least able to sustain it. Similar facts were brought out in other states, and state aid for rural schools was legislated in several states. It seemed anomalous that extensive road systems could be built while public schools were being closed, but this was primarily due to the fact that roads were built by the state from funds obtained from various forms of taxation, whereas in many states the schools were almost entirely supported by local taxes on real estate.

An increased burden was placed on rural high schools by the larger number of students and postgraduate students resulting from a curtailment of the usual rural-urban migration. In some

---

[11] Brunner, E. deS. and Lorge, Irving. *Rural Trends During Depression Years 1930-1936.* New York: Columbia University Press. 1937

[12] For the best brief summary of rural school problems see "The Outlook for Rural Education." *Research Bulletin of the National Education Association.* Washington, D.C. Vol. 9. No. 4. September 1931. See also Renne, Roland R. "Rural Educational Institutions and Social Lag." *Rural Sociology.* 1:306-321. September 1936

districts, more particularly those within driving distance of cities, the open-country schools had a considerable increase in pupils on account of the migration from cities.

On the other hand, there were some definite gains achieved as a result of these conditions. Thus, in spite of lowered salaries, competition for teaching positions was keener and there was a notable improvement in the training of rural teachers. Dr. Brunner[13] reports that the proportion of white teachers with college degrees in open-country schools rose from 4.5 per cent in 1924 to 21.7 per cent in 1936. It will be interesting to study the effect of better trained teachers on individual schools and whether they will make the schools more of a force in the life of rural communities or whether they are coming from the towns and cities and going home week ends and not identifying themselves with the community life.

As a means of reducing costs rural school consolidation was hastened in some states and in a few cases may have been carried to an extreme. At least this is the opinion of various persons familiar with local situations. For example, a correspondent from West Virginia writes:

> The depression brought about two amendments to our state constitution which limited levies on farms and small homes. This forced school consolidation to its final conclusion in a very rapid manner. In some sections this consolidation has been disastrous to community life with the school children riding 10 to 20 miles in buses to attend their classes. As yet we do not know what will happen to community life in many sections of the state.

Too rapid consolidation, and the consolidation of large areas for purposes of "efficiency" and "economy," without considering the effect on rural community life, have had an unfortunate effect, as is further illustrated by the following from a Louisiana correspondent:

---

[13] Brunner, E. deS. and Lorge, Irving. *Op. cit.* P. 157

In some parishes children go as far as thirty or thirty-five miles by bus, after walking three or four miles and ferrying across a river to catch the bus. Sometimes they must leave before daylight and return after dark. In extra-curricular activities discrimination against the country children is common because they cannot remain after school hours to participate in such activities. In other words, the school has become something apart from the lives of the people in the local communities. It is merely a place where the children go five times a week for formal instruction.

Surely such economy does not produce real efficiency when measured in terms of the best interests of the pupils or of the communities.

On the other hand, in some of the rural states more acutely affected by drought, consolidation seems to have declined. Thus in North Dakota it is reported: "The consolidated school movement, so aggressive two decades ago is making no gain at the present time. High cost of transporting pupils and other operating expenses have exceeded the districts' capacity to pay. Especially in the western half of the state, the sparseness of the population and the financial situation of the districts have shown a complete transportation system to be impractical and in several counties transportation as a policy has been abandoned."[14]

The growth of consolidation is shown in the larger porportion of country children in village schools. Thus, Dr. Brunner[15] found that in 1936, 29.2 per cent of the pupils in the village elementary schools were from the country, whereas in 1930 this was 25.7 per cent and in 1924 it was 24 per cent, showing a much more rapid gain in the last six years. The proportion of open country children in village high schools also showed a definite increase. This growth of village schools necessitated many new buildings or additions to old buildings. Although there was less building

---

[14] McCrae, J. A. *Survey of Rural Education in North Dakota.* Federal Emergency Relief Administration for North Dakota. 1935

[15] Brunner, E. deS. and Lorge, Irving. *Rural Trends During Depression Years, 1930-1936.* New York: Columbia University Press. 1937

during the early part of the industrial depression, a large number of new schools and additions which otherwise would not have been constructed for a number of years were built with the assistance of WPA funds. As a result Dr. Brunner finds that in 128 of 140 villages new schools or large additions had been built since 1924.

This whole trend of country children into village schools raises the question of how this has affected village-farm relationships. Has it increased community solidarity or has it aroused antagonism on the part of farm people? In the cases quoted above the answer is clear that community solidarity has been injured, and where the consolidations have covered too large areas it is doubtful whether they will ever be beneficial to community life. On the other hand, where centralization has been brought about as the result of the desire of the people in the area, it has given farm people a representation in the control of the high school and has undoubtedly resulted in making the consolidated school a major force in community organization. It is highly desirable, therefore, that at the present time the development of consolidation be studied in many rural communities with a view to ascertaining the effect of school activities upon community organization. The process of consolidation is still young and will doubtless increase in the future. It is important that the rôle of the school in community life be clearly analyzed, so that pupil costs and supposed efficiency of over-large schools may not result in losing the interest and support of the farm patrons and weakening the life of the rural community. If community values are considered the centralized school may be the means of creating a richer community life.[16]

One important fact revealed by the studies of relief families

---

[16] For a very strong affirmation of this view by a leading educator, see Moehlman, Arthur B. *The Community School District.* University of Michigan School of Education Bulletin. 7. No. 4. Pp. 49-51

in rural areas is the high proportion of illiteracy and lack of schooling among them, which undoubtedly is related to their dependency, and which limits the possibility of their rehabilitation. This is particularly noticeable in the South. In the counties studied in the survey of rural problem areas, "one-half of the Negro family heads and one-fifth of the whites in the Eastern Cotton Belt reported no schooling, and four-fifths of the Negroes and about one-half of the whites had less than five years. Although the percentage of family heads with no schooling in the Appalachian-Ozark Area was less than for whites in the Eastern Cotton Belt, the proportion that had completed fewer than five grades (56 per cent) was larger."[17]

The effect of inadequate schooling is clearly revealed by a statement made by Dr. W. D. Nicholls, head of the department of farm economics at the University of Kentucky, who has been studying the movement of families from the mountain counties of that state: "There was during the depression a very heavy return of families that had moved out of the mountain areas to such industrial centers as Youngstown, Detroit, and Akron, when the family bread-winners lost their jobs. The great majority of these men had a very meager education and this seems to have been a factor in causing them to be among the first laid off when the depression struck the industrial plants in which they were employed."

The difference in education between families receiving relief and their nearest neighbors was brought out in a survey made of 47 counties representing the 13 major agricultural regions of the United States in October 1933.[18] "Nearly 8 per cent of all relief heads surveyed had never attended school [5.4 per cent white, 28.3 per cent Negro], in comparison with 3 per cent

[17] Beck and Forster. *Op. cit.* P. 90

[18] McCormick, Thomas C. *Comparative Study of Rural Relief and Non-Relief Households.* Washington, D.C.: Works Progress Administration. Division of Social Research. Research Monograph II. 1935. P. 30

of the heads of households not receiving relief [1.5 per cent white, 25.2 per cent Negro]. An additional 19 per cent of the relief [17.6 per cent white, 36.9 per cent Negro] and 11 per cent of the non-relief heads [9.1 per cent white, 44.5 per cent Negro] had not progressed as far as the fifth grade, having achieved little more than the bare ability to read and write. Less than half of the heads of relief households [49.3 per cent white, 11.7 per cent Negro], compared with two-thirds of their self-supporting neighbors [70.5 per cent white, 8.2 per cent Negro] had completed grade schools or better." The clear picture of the effect of illiteracy on Negroes and of the influence of schooling on social changes, given by Johnson[19], is worthy of study in this connection, as suggestive of promising lines of research.

It would seem desirable to study the relation of the education of the heads of households to their ability to become self-supporting from the records of families being rehabilitated by the Resettlement Administration. It may also be possible to show whether heads of relief families with more schooling were able to get off relief rolls sooner than those with less education, by a study of the surveys made by the FERA of opened and closed cases. It should also be possible to determine whether there is an increase in the proportion of illiteracy in the heads of families who remain on relief as the number of relief families decreases.

Certainly if any permanent improvement is to be effected in the condition of the poorer tenants and croppers it will be impossible by any means unless they are compelled to obtain much more schooling than at present.

Other important changes in rural schools may have been hastened by the depression, such as the more general introduction of social science courses in the high schools, the increase in vocational courses in agriculture and home economics, and

---

[19] *Op. cit.* Chapter IV. (See note 6, p. 81)

the introduction of vocational guidance, but these are more appropriately within the field of educational research.

One important effect of the difficulty which many of the poorest areas have experienced in maintaining their schools, particularly with regard to high schools, has been a definite tendency to question the public responsibility for secondary education and the rôle of the school in the community. Thus Mr. McCrae summarizes the situation in North Dakota:

> Rural Education in North Dakota is passing through a critical stage in its development. Critical, not alone by reason of the crop failure and low prices for farm commodities which have impoverished the country and have checked the flow of revenue into school treasuries, but because of a state of restless uncertainty in the public mind as to the place which the public schools rightly hold in the social structure.[20]

Dr. Brunner cites one instance of a similar tendency: "In one village the taxpayers' league discovered that almost one-half the parents who had two-thirds of the children had started an agitation to exclude the children of non-taxpayers from the schools unless they paid tuition."[21] In most of the country there seems to have been a definite increase of interest in the schools and an appreciation of their enlarged services to the community, but in some areas, because of the severity of the financial situation, the opposite result seems to have occurred. This is a difficult matter to evaluate, but evidence might be obtained from a study of local newspaper editorials and through interviewing county school superintendents and high school principals. It will be important to locate these areas of discontent, because they may be able to block desirable changes in school legislation, and also they may reveal the need for state and federal aid to meet their situation.

---

[20] McCrae, J. A. *Survey of Rural Education in North Dakota.* Federal Emergency Relief Administration for North Dakota. 1935. P. 41

[21] Brunner, E. deS. and Lorge, Irving. *Rural Trends During Depression Years, 1930-1936.* New York: Columbia University Press. 1937. P. 144

ON THE RURAL CHURCH

The effect of the depression on the church throughout the nation will also be discussed in another monograph of this series,[22] so that only certain aspects of the situation of the rural church will be considered here.

There is little factual evidence with regard to any considerable change of status in the rural church caused by the depression. Correspondents in widely scattered states unite in their opinion that there has been a decline in the number of open-country churches and that more farm people are attending village churches. This movement has been going on for the past ten or fifteen years and has doubtless been hastened by inability to maintain the small country churches. On the other hand, well-informed church administrators state that many small country churches which had been closed or would have soon closed, were revived so as to give a living, however meager, to unemployed ministers. This tendency has had an unfortunate effect on the larger parish plan which was just getting well under way and had aroused general interest as to its possibilities, when it was halted by the industrial depression. Rev. Mark A. Dawber, the superintendent of rural work of the Methodist Episcopal Church, says concerning it:

During the recent years of the depression the larger parish plan has been having a difficult experience due to several causes, the primary one being that of having a larger number of ministers than the several denominations could absorb in a time of retrenchment. Naturally the several administrators and supervisors have been reluctant to move in directions that would reduce the number of parishes. The result of this is an increasing number of pastors on starvation salaries, the continuing of competitive churches, the struggle to make ends meet, and the lack of a vital program of diversified activities to challenge the life of the communities. This may only be temporary, at least we hope so, but in any case, the larger parish principle must not be allowed to die.[23]

---

[22] See Kincheloe, Samuel C. *Research Memorandum on Religion in the Depression.* (monograph in this series)

[23] Dawber, M. A. "The Rural Church Today." *The Rural Church Today and*

In many cases there has been renewed interest in the uniting of rural churches by federation or otherwise in overchurched communities. As a result of a statewide survey in Missouri, Sneed and Ensminger[24] state:

> From the foregoing it appears that there is a rather definite union church trend in Missouri. It is not clear, however, whether it is a somewhat permanent development or whether it is a temporary one brought about in the present emergency largely for the purpose of economic convenience.

The rise of small churches of emotional, pietistic sects is noted in one or two states. One correspondent reports that the idea that the drought and the depression are punishment for human wickedness is fairly common and points out that religious emotionalism might lead to erratic political behavior as a means of obtaining release from tensions, but no concrete examples of this trend have been reported. In general there seems to be no evidence that in rural communities there has been any general revival of religion, which has often been held to be a concomitant of hard times.

The only comprehensive study of the rural church covering the depression period for a whole state is that made by Sneed and Ensminger in Missouri, in which they surveyed some 3,000 rural churches in all but four counties. For the period 1929 to 1934 they found[25] that the total rural church membership in 1,035 churches making reports on membership increased 4.08 per cent, while for the period 1920 to 1934 it increased 5.46 per cent, indicating a larger gain in the later period. They also

---

Tomorrow: A Report of the National Conference on the Rural Church. New York: The Home Missions Council. 1936. P. 12

[24] Sneed, Melvin W. and Ensminger, Douglas. *The Rural Church in Missouri.* University of Missouri Agricultural Experiment Station. Research Bulletin 225. June 1935. P. 62

[25] *Op. cit.* P. 16

indicate that the gain was largest in the larger villages and least in the open-country churches.

Dr. B. Y. Landis, of the Research Department of the Federal Council of Churches of Christ in America, states[26] that "only one body, the Congregational and Christian churches, has available statistical studies of church attendance covering a period of years." He shows that for this denomination as a whole there was no noticeable change in attendance at Sunday morning services between 1930 and 1933. It would be valuable to separate the attendance, and also the contributions, of rural congregations of this denomination from those of cities in the same states to determine any differences.

In three counties in Wisconsin Kirkpatrick[27] found that church attendance decreased only 10 to 15 per cent although contributions decreased 20 to 38 per cent in 1933 from 1929. This indicates that in these areas interest in the church had been fairly well maintained, for attendance would tend to decline on account of less ability to operate automobiles and to maintain customary standards of clothing.

Mr. Guy T. Gebhard, Town and Country Y M C A secretary at Wichita, Kansas, has kindly furnished data from a survey made by him of 47 rural Protestant churches in Sedgwick County, Kansas, comparing 1920, 1930, and 1935. He finds that the total membership of these churches increased 6 per cent between 1920 and 1930, and 14 per cent between 1930 and 1935, while the membership of boys and girls under 21 decreased 6 per cent in the first decade but increased 64 per cent in the last five years. Similarly membership in the young people's societies decreased

---

[26] Landis, Benson Y. "The Church and Religious Activity." *Am. Journal of Sociology.* 40:783. May 1935

[27] Kirkpatrick, E. L., Tough, Rosalind, and Cowles, May L. *How Farm Families Meet the Emergency.* Madison: University of Wisconsin Agricultural Experiment Station. Research Bulletin 126. January 1935. P. 27

22 per cent between 1920 and 1930, but gained 19 per cent from 1930 to 1935. The total budgets increased 26 per cent from 1920 to 1930 but declined 42 per cent in the last five years; while benevolences declined steadily, 39 per cent from 1920 to 1930 and 49 per cent from 1930 to 1935. Benevolences per active member in 1935 were only about one-fourth what they were in 1920, and the total cost per active member was about one-half as much in 1935 as in 1930. But attendance at Sunday morning services increased 25 per cent from 1930 to 1935. All of this indicates an increased interest in the church, in spite of poorer support, in this individual county.

The only nationwide study of changes in the rural church during the industrial depression is that made by Brunner and Lorge[28] in their resurvey of 140 agricultural-village communities in 1936. This gives a less encouraging view than that of the surveys in Missouri and Kansas mentioned above. Attendance at church services, which is probably the best index of interest, declined about one-fifth, which was three times as large a decline as in the previous six years, and this was largest in the Middle Atlantic and Midwestern states. They find that the decrease in the number of rural churches was less than half of that in the previous 6 years. Although 20 per cent of the open-country churches of 1930 had closed, there were only 8 per cent fewer village churches, and four-fifths as many new churches had been started, so that the net loss was only about 3.4 per cent. This study confirms the suggestion previously made that many rural churches were maintained or reopened in order to furnish a living to ministers who otherwise would have been unemployed. Although there was an increase in the membership per church, resulting largely from the smaller number of churches, the ratio of church members to the population showed a loss in the Middle Atlantic and Midwestern states, but a gain

---

[28] Brunner, E. deS. and Lorge, Irving. *Rural Trends During Depression Years 1930-1936.* New York: Columbia University Press. 1937. Chapter XII

in the South and Far West. There was a definite gain in church consolidation, but no striking advance in this movement as might be expected from motives of economy. It should be noted that this survey covered villages and their communities in which the villages averaged over 1,300 population, so that it is not necessarily representative of the smaller village communities which form the large majority.

It would be expected that the financial support of the rural church would be seriously affected by the depression. This was the more necessary because the cost of rural churches had been steadily rising ever since the World War and there had already been a reduction in the amount contributed to church benevolences before the industrial depression. Thus from the data of the federal census, Dr. C. Luther Fry[29] states that the churches in town and country areas in 1916 spent $8.40 per member, which increased to $16.41 per member in 1926. In one of the poorer rural counties of southern New York Mather showed that the average financial load per member in 1930 had increased three times since 1900 and had doubled since 1915.[30] The evidence on this point for Missouri is clearly revealed by the study of Sneed and Ensminger, who show that although the current expenses of rural churches increased about 10 per cent from 1920 to 1929, they declined 27 per cent from 1929 to 1933, during which period salary expenses declined 29 per cent and benevolences declined 38 per cent. "In 1933 the total sum of subscriptions and collections for the total rural area was 32.65 per cent less than the total in 1929."[31] It is interesting to note that this decline was greatest in the villages of 1,000 to 1,500 population and least in the villages of under 200. This

---

[29] Fry, C. Luther. *The U. S. Looks At Its Churches.* New York: Institute of Social and Religious Research. 1930. P. 96

[30] Mather, Wm. G., Jr. *The Rural Churches of Allegheny County.* Ithaca: Cornell University Agricultural Experiment Station. Bulletin 587. March 1934. Table 5, p. 8

[31] *Op. cit.* Pp. 58-60

agrees rather closely with the findings of Dr. Brunner's[32] survey, which showed that per capita contributions dropped one-third (this decrease being slightly less in the open-country than in the village churches) while the contributions for benevolences in the village churches declined by one-half from 1930.

Studies of the statistical reports of local conferences, synods, and associations of the various denominations for the period 1925 to 1935 might reveal trends in the decline or growth of membership and financial support, but are not very satisfactory for getting any clear picture of the number of churches closed or opened. About the only method of determining the latter point is to make a first-hand study of the churches opened or closed in a given county for the past ten years. This might well be made from 1926 to 1936 to correspond with the dates of the Census of Religion of the U. S. Bureau of the Census. The cooperation of state and county federations of churches might well be enlisted in making such county surveys, which would involve but little expense if carried on by local ministers. Care should be taken to include all churches of all sects. That such surveys are entirely feasible is shown by the excellent report on the Missouri situation, which was made as a Civil Works Administration project in 1934, and which offers many suggestions for studies of this kind. It is particularly valuable in that it classifies the data according to the size of the community, but it would also be desirable to make tabulations for sections of different economic status or types of agriculture. Studies of this sort might well be made of the year 1936 so as to check with the federal census on religion, and it would be desirable, in so far as possible, also to obtain records for 1933 as the low point of the depression, and for 1929 or 1930 as a base.

ON THE RURAL COMMUNITY

All of the institutions and factors which have been discussed above have their setting in the rural community. Has the de-

[32] *Op. cit.* Chapter XII

pression changed any of the trends of community life? Inasmuch as a rural community area consists of a village and the tributary farm territory, the integration of the interests of these two elements is an emergent process. It has been suggested that one effect of the depression has been to cause villages to realize their dependence upon the farmers, both for business and for the support of village institutions, as never before. In the past there has been considerable friction between farmers and villagers, but good roads and automobiles have made farmers more independent of the local village and, in consequence, village people have been forced to seek their patronage and to give them a share in the control of village institutions and organizations.[33] In some cases, however, the depression may have developed new causes of conflict. In his resurvey of villages Dr. E. deS. Brunner[34] found 13 cases of conflict in 1935 as against 6 in 1930, although he also reports that village-country relations had improved in over one-third of the 140 villages studied. His field workers reported a general increase of understanding between farmers and villagers, and that the latter were keenly aware of the effect of the farmers' AAA checks upon village business. These were mostly larger villages (average size 1,351 persons in 1930) and it would be desirable to study the relations in the more numerous smaller villages.

According to Dr. Brunner's survey, the previous tendency for country neighborhoods to decline in number has continued and has been hastened by the building of more farm-to-market roads. He found that 34 per cent of the neighborhoods active in 1930 had disappeared or were inactive in 1936, but some new ones had arisen or revived, so that there was a net loss of 23.5 per cent. It is probable that the decline of neighborhoods would

[33] On this point see the conclusions of Clark and Roberts concerning the situation in Kansas. People of Kansas. Topeka: Kansas State Planning Board. October 1936. P. 184

[34] Op. cit. P. 87. (See note 28, p. 94)

strengthen community life, although this is by no means inevitable.

If neighborhoods have been affected by more good roads to market villages, is it not probable that small villages and hamlets have been similarly affected? Although most incorporated villages have grown about the same as the total population,[35] it is also true that the smaller incorporated villages have shown an increasing tendency to decline in population during the present century.[36] The only considerable body of evidence to show whether smaller unincorporated villages are also tending to decline more rapidly is that recently presented by Clark and Roberts for the state of Kansas, based on census data and populations reported in commercial atlases, for 1910 to 1930. They conclude:

> In the state as a whole, the larger size-groups have a larger per cent of towns that are growing and a smaller per cent that are declining in population than the smaller size-groups. The cities head all classes in the 'growing column,' the large villages come next, followed by the medium-sized and small villages in order. The hamlets seem at first to be an exception, since a larger proportion of them are growing than of the small villages, but they are also declining in greater proportion. The small unit (25) applied in the measurement of hamlets is mainly responsible for this apparent deviation from the general tendency.[37]

This study is the more valuable because it separates three regions of the state differing in density of population. Studies made by Lively[38] in Minnesota and by Smith[39] in Louisiana,

---

[35] Brunner, E. deS. and Kolb, J. H. *Rural Social Trends.* New York: McGraw-Hill Book Co. 1933. Chapter III

[36] Gillette, John M. *Rural Sociology.* New York: Macmillan Company. 1936. 3rd ed. P. 580

[37] Clark, Carroll D. and Roberts, Roy L. *People of Kansas: A Demographic and Sociological Study.* Topeka: The Kansas State Planning Board. 1936. P. 183

[38] Lively, C. E. *Growth and Decline of Farm Trade Centers in Minnesota, 1905-1930.* Minneapolis: University of Minnesota Agricultural Experiment Station. Bulletin 287. Table 10, p. 12

[39] Smith, T. Lynn. *Farm Trade Centers in Louisiana, 1901 to 1931.* Baton Rouge: Louisiana State University Agricultural Experiment Station. Bulletin 234. P. 29

using the number of business units as a measure of growth or decline, also indicate that smaller unincorporated villages were declining more rapidly up to 1931. It will be desirable, therefore, to determine whether the depression has hastened this trend for the smaller unincorporated villages to decline, or whether the additional population resulting from immigration and the stopping of emigration has enabled them better to maintain their position in competition with the larger villages and towns. As the census does not enumerate unincorporated villages and it is doubtful whether the estimates in commercial atlases are sufficiently accurate for so short a period, the only method of determining this matter would seem to be by a study of the number of business units as reported in Bradstreet's rating books or as obtained by a personal survey. Lively has shown that there is a very close correlation between growth or decline of population with that of the number of business units, so that this may be sufficiently accurate for practical purposes, where definite enumerations of population are lacking.

One evidence of increased interest in the rural community is an indigenous movement in southern Virginia, during the last three years, of Ruritan Clubs, the preamble to whose statement of objectives recites "that the greatest handicap to the American farmer and the rural communities is the inability of the rural and village folk to get together regularly where problems affecting the rural life of the community, the State, and the nation, may be systematically discussed and thoroughly considered." One of the main objectives of these clubs is "to unify the efforts of individuals and institutions in the community, in making it a better place in which to live."

In Wisconsin, in three areas previously studied by him, Kirkpatrick reports:

Those clubs with major emphasis on the educational and social aspects of community life seem to be 'holding on' if not gaining ground. In addition, those community groups (P.T.A., Ladies' Aid and the like) which sponsor their own programs suited to local needs are functioning more actively and effectively than the groups which are satisfied to carry programs sent out from central (district and state) offices. Apparently attempts to meet local situations and take care of local needs call for ingenuity, spontaneity, and ability, which when used cooperatively result in worthwhile accomplishments even in times of emergency.[40]

In two or three states, particularly in the South, the county-wide organization of relief administration and of other federal administrations has centered interest in the county seat town and seemingly has so affected the other villages as to cause definite resentment and complaint from merchants in these villages. In New England especially, there appears to be a feeling that federal and state relief administration has tended to diminish the responsibility of the town and that town administration was better for maintaining local interest and community morale. The too rapid consolidation of schools or the creation of over-large districts may have hurt the smaller communities, as has been indicated in the discussion of the school situation (pages 83-90).

These are some of the possible adverse effects of the depression on community life. Whether they are general or whether the increased interest in community organization and a larger tendency to pull together are equally or more common, we have no means of knowing. Nor does there seem to be any source of information on this point other than a personal survey of a considerable number of communities in areas of various types. Such surveys should be made by persons who have had experience in community studies and in judging the reliability of information.

It would seem that the best way of ascertaining the effect of the depression on community life would be to interview editors of local newspapers, ministers, bankers, school principals, leading merchants, county agricultural and home demonstra-

---

[40] Kirkpatrick, E. L., Tough, Rosalind, and Cowles, May L. *How Farm Families Meet the Emergency.* Madison: University of Wisconsin Agricultural Experiment Station. Research Bulletin 126. January 1935. P. 35

tion agents and others who are in positions of natural leadership, for it would be quite difficult to obtain now much quantitative data of any value about community conditions before and during the depression. Only for communities which have been previously studied, such as the villages resurveyed by Dr. Brunner, or such as Orange township, Blackhawk County, Iowa, studied by Dr. VonTungeln, would quantitative data be available for the predepression and depression periods. It would, of course, be possible to ascertain the number of churches, schools, and business firms, and other local organizations which existed in the communities in 1930 and the number that exist at present, with the dates on which they started or closed, and thus get a rough measure of the effect of the depression on changes in the organizational patterns of the communities.

This brings out the desirability of recurrent surveys of a series of rural communities which are fair samples of different areas. Just how to determine the validity of the sample selected is a matter which needs thorough analysis, for there is as much difference in the character of communities as in individual persons. Particularly is there need for the repeated study of the smaller rural communities whose village centers have from 200 to 500 inhabitants, for they form the vast majority and are the ones which are being most affected by competition with larger centers.

SUMMARY

There seems to be little evidence that the depression has had any fundamental effect toward changing the patterns of the basic rural institutions, although certain tendencies already under way have been accelerated while others have been retarded. Attention has, however, been directed to certain problems of these institutions which warrant further investigation, among the more important of which are: (1) surveys of rural housing in the South; (2) studies of changes which have resulted in farm family relationships because of more children remaining at home; (3) the effect of rural school consolidation in the depression on the life of smaller rural communities and on community organization; (4) the relation of the amount of schooling to the ability of rehabilitation clients to become self-supporting; (5) county and state surveys of rural churches opened and closed during the depression; (6) changes in the number of business concerns in small villages during the depression; (7) the desirability of recurrent surveys of rural communities typical of different sections and areas to show what changes have occurred.

Chapter VII

Effects of the Depression on Rural Services

During the last two decades there have been developed several classes of services which supplement and support the basic rural institutions discussed in the last chapter. Among the more important are the public health service, the extension service in agriculture and home economics, social welfare work, and the movement for an organized program of recreation. The influence of the depression on these services and the need for research concerning any changes which may have occurred, are considered in this chapter.

ON HEALTH SERVICES AND FACILITIES

Various studies have been made of the effect of the depression on health, and the incidence of particular diseases, particularly with regard to malnutrition of children and concerning morbidity rates, but research on this topic is within the field of public health experts and requires techniques which only those with medical training are qualified to use.[1] Although social scientists will wish to use the results of such studies, their own concern is chiefly with the extent to which health services and facilities have been affected by the depression and whether rural people show more or less interest in them.

There seems no logical reason why there should be any de-

---

[1] See Collins, Selwyn D. and Tibbitts, Clark. *Research Memorandum on Social Aspects of Health in the Depression.*

crease in the number of rural physicians as a result of the depression, except possibly in parts of the western plains most seriously affected by drought; but there would be reason to suppose that recent graduates of medical colleges, finding less practice among the unemployed in the cities, might seek to establish themselves among a rural clientele. This hypothesis is worthy of testing by a study of the medical directories for different states. Dr. H. G. Weiskotten, dean of the College of Medicine of Syracuse University, who has been making studies of the graduates of medical colleges for the past twenty years at five-year intervals, tends to support this hypothesis. He writes that: "There has recently been a definite movement of the young graduates to the rural areas. My data suggest that 24.9 per cent of the 1930 graduates are now located in communities of less than 5,000 as compared with 18.1 per cent of the 1925 graduates in 1931. The total number dealt with in the last study is larger than in the previous studies so that the absolute number locating in the rural area is even greater than the percentage difference indicates." Dr. Brunner reports that the number of physicians has increased in the 140 villages which he studied.

County health units had increased rapidly prior to the depression, growing from 10 in 1915 to 81 in 1920, 268 in 1925, and 504 in 1930. The maximum was reached in 1931 when "health units were in operation in 573, or approximately one-fifth, of the counties in the United States, and they served a population of 25,074,278, or about one-fifth of the total population of the country."[2] In 1932 and 1933 the number declined to 519. Since then no figures have been published, but the author has been informed by the Public Health Service that, as a result of the Social Security Act, federal grants for this work were augmented in 1936 and that the work has been greatly extended.

[2] Ferrell, John A. and Mead, Pauline A. *History of County Health Organizations in the United States, 1908-1933*. Washington, D.C.: U. S. Public Health Service. Public Health Bulletin No. 222. March 1936. Pp. 32, 33

Although the number of county health units temporarily decreased, the work of the state and federal relief administrations has compelled them to study the health needs of their clients and to see that adequate health services were made available. This resulted in the employment of many public health nurses by the relief administrations in counties which had not previously had their services, and may be the means of demonstrating the need of continuing them under local auspices. In several states dietitians employed by the relief administration, or employed by the extension services or the Red Cross and cooperating with the relief administration, rendered important service in the improvement of nutrition. As a result, a better knowledge was gained of the food habits of the poorer classes of rural people, which should have a permanent effect upon programs of work with them. A definite advance was made in many localities, particularly in the South, in the use of home gardens, thus increasing the vegetable diet. As a means of furnishing work, many thousands of sanitary privies were constructed and distributed throughout the South, which, it is to be hoped, will have a lasting effect on rural sanitation.

However, among the mass of the rural people who were able to maintain independence from relief, there was probably less use of available health services because of lack of funds. Thus, in three localities in Wisconsin, Kirkpatrick[3] found that family expenditures for health purposes in 1933 had decreased from 35 to 50 per cent from those of 1929.

Possibly because of this inability to pay for health services, observers familiar with local conditions in several states inform the writer of an increase of interest in public health work, and one or two mention interest in mutual or cooperative associations for the employment of physicians and the maintenance

[3] Kirkpatrick, E. L., Tough, Rosalind, and Cowles, May L. *How Farm Families Meet the Emergency.* Madison: University of Wisconsin Agricultural Experiment Station. Research Bulletin 126. January 1935. Pp. 13, 14

of small local hospitals. Health conditions have become so acute in some states that experiment station directors have been convinced that rural health problems need study and in several cases members of their staffs are actively at work on them.

The determination of definite gains or losses in rural health administration and practices will require repeated inventories of local conditions in communities or counties. For rural communities, Dr. Henry J. Burt,[4] with the cooperation of the Missouri State Board of Health, developed a "health index" for measuring the changes in given communities, and the American Public Health Association has an Appraisal Form for Rural Health Work for counties. For more comprehensive research proposals in this field the reader is referred to the monograph in this series on the social aspects of health.

ON RECREATION AND PLAY[5]

The opinion of rural leaders from all parts of the country is uniform that there has been a notable increase of interest during the depression in non-commercial recreation in which the people themselves participate. This trend is but natural, both on account of lack of funds for commercial amusements and because of the increased number of young people in the rural communities. A rural leader in Maryland, who has had wide experience in other parts of the country, writes:

I am of the opinion that farm people, by and large, are redeveloping some features of a distinctive rural culture. Possibly this shows up most clearly in the realm of recreation and leisure time activities. Home-talent performances, musical activities, athletics, and other forms of participator recreation have become more popular in a number of rural areas

---

[4] Burt, Henry J. *Rural Community Trends.* Columbia: University of Missouri Agricultural Experiment Station. Research Bulletin 161. October 1931. P. 27; *Rural Community Trends: Second Report.* Research Bulletin 199. June 1933. P. 42

[5] See also Steiner, Jesse F. *Research Memorandum on Recreation in the Depression*

which I have known intimately for many years. Somewhat less pronounced but none the less significant, is the revival of various arts and skills in the home and on the farm. Some of these are used as means of reducing former cash expenditures; others represent new contributions to living.

Amateur drama, folk dancing, square dances, and singing games are reported to have greatly increased in popularity. A Louisiana correspondent writes that fish-fries, picnics and family reunions have been considerably revived. However, as the depression lifts there seems to be a tendency to revert to commercial amusements—moving picture theatres and others. Sunday movies are said to be increasing again in villages and towns in Iowa.

This movement has been greatly stimulated by the extension services of the agricultural colleges, several of which have employed state leaders for recreation in the last few years. The WPA has also employed recreation leaders who have worked with the extension services and with the schools. As a result of this and previous work, there is coming to be a definite demand for this sort of service in rural communities. At the conference of farm women held in Washington, D.C., in the fall of 1934, one farm woman from northern Minnesota stated that they should work to obtain an employed recreational leader in every county, and another urged that all county extension agents, both men and women, should be given training in recreation during the college course as a part of their necessary equipment for extension work. Such statements, though made by the more enthusiastic, show a very definite advance in public sentiment with regard to the importance of recreation and play in rural life.

Facilities for recreation have been extensively improved by WPA projects and the work of CCC camps, in the development of rural parks and picnic grounds, out door swimming pools, the improvement of riversides, hilltops and other scenic spots, and the building of mountain trails. In some states, particularly in the South, there has been some development of outdoor recreation centers. As a result of this interest in local recreation, several county councils on recreation have been started in New York State, for promoting local programs and for mutual assistance in training local leaders. These include representatives of various organizations and agencies concerned with recreation. Rural high schools and consolidated schools have given much more attention to recreational projects for those out of school and even for adults, particularly where they have had the help of WPA recreation leaders.

So far as rural life is concerned it would seem hardly worthwhile to attempt any quantitative measurement of recreational activities as affected by the depression. It would be desirable to ascertain through community studies to what extent schools and churches have any definite program or policy with regard to recreation, and whether these and other organizations have any coordination of effort in this field or are in free competition with each other. Recreational activities have very definitely been fostered during the depression as a means of maintaining morale. Valuable studies in social psychology might be made of how participation in recreation affected the morale of individuals and families and what types of activities seemed most successful for this purpose.

ON EXTENSION WORK AND ADULT EDUCATION[6]

During the depression, the extension work in agriculture and home making conducted by the state colleges of agriculture and the United States Department of Agriculture demonstrated its worth to farm people as it did during the World War, and made itself so invaluable that it is more firmly established than ever.[7] Beset by drought, low prices, and insect pests, farmers

---

[6] See also Educational Policies Commission. *Research Memorandum on Education in the Depression.*

[7] For details see *Report of Extension Work in Agriculture and Home Eco-*

turned to the county agricultural agents for advice as to how to meet their difficulties, and without the aid of these agents and their organizations the federal government would have found it almost impossible to administer the Agricultural Adjustment Act and other federal legislation. They were the greatest assistance to the Farm Credit Administration and to the relief agencies. Indeed, the work connected with the administration of federal legislation, which was thrown upon them, came to require the major part of their time in many parts of the country and the usual educational program was slowed up accordingly to meet the emergency.[8] This was inevitable and it was fortunate that so well organized a system, having the confidence of the farmers of the country, had been built up. This does raise a question, however, as to whether, if this sort of thing should go on over a series of years, it would not seriously change the relation of the agents to the farmers by making them exponents of policies of the federal administration.

One notable effect of the depression is that it has forced the county agents to give more time to working with the poorer class of farmers, with those on relief or who have needed loans of one sort or another. Inasmuch as the agent is seeking to introduce and establish better methods of agriculture, it has been inevitable that his main constituency has been the more intelligent and progressive farmers who would make use of this aid, and these have commanded so much of the agent's time that he has tended to neglect what may be called the marginal farmers who most need help but are the hardest to interest.

nomics in the United States, 1931. Washington, D.C.: U. S. Department of Agriculture. Extension Service. Mimeo. (also for 1932) ; and "Serving American Agriculture." _Report of Extension Work in Agriculture and Home Economics._ 1933. Mimeo.

[8] Thus the official report of the cooperative extension work shows that the county agricultural agents used 36 per cent of their time on AAA work and 5 per cent on relief work, and only 59 per cent on regular extension work in 1935. This being the average the figures would be much higher for many counties.

Although this work was mostly taken over by the rehabilitation agents of the FERA and now by the local agents of the Resettlement Administration, the county agricultural agents were constantly called upon for advice and active assistance because of their knowledge of local conditions and people, and the work of the Resettlement Administration is intimately coordinated with that of the extension services. Indeed, it is a question whether most of the follow-up supervision of resettlement clients will not fall upon the county agricultural agents.

Another effect of the depression has been to greatly strengthen the educational program with regard to the marketing of farm products and the economic aspects of agriculture. With additional federal funds for this purpose, there have been intensive educational campaigns, the results of which are seen in the very general demand of farmers for the stabilization of the value of money and in their support of the AAA. Indeed, it has been suggested that in some localities the farmers are much better educated on economic problems than are the local business men. Certainly no such campaign of education on economic problems has ever before been conducted by men trained in economics, among so large a class of our population, and its results will inevitably be far reaching.

The work of the home demonstration agents was equally important in showing housewives how to reorganize their family budgets, how to conserve food by canning and storage, how to remodel clothing and furniture, to produce gardens, and to conserve health and morale. In hundreds of counties they were called upon for advice and assistance by the local relief administrations, and in many areas they took part in organized campaigns for improving the diet and health of relief clients. Thus,

See Wilson, M. C. *Statistical Results of Cooperative Extension Work, 1935.* Washington, D.C.: U. S. Department of Agriculture. Extension Service Circular 244. June 1936. Pp. 52. Mimeo.

they have also been compelled to give much more attention to the poorer class of rural people with whom they have previously had less contact.[9]

One of the most interesting phenomena of the whole agricultural depression has been the progress of the junior extension work with rural boys and girls, commonly called the 4-H Club work, which has shown a steady growth since 1915, excepting very slight decreases in membership in 1933 and 1934, until it now enrolls over a million rural boys and girls.[10] This is undoubtedly due not only to the popularity of the work, but to the fact that it involves little expense and in most cases brings the boy or girl some cash income.

Partly to retain the interest of older youth who dropped out of 4-H Club work and partly to meet the needs of the increased number of older youth in rural communities, the extension services—national and state—have become interested in developing extension work and some form of organization for older youth,

[9] The need for more aid to marginal families by extension workers has become more apparent particularly in states having a considerable proportion of subsistence farms. Thus the sociologist of the Virginia Agricultural Experiment Station writes:

"Among the many measures needed for dealing constructively with these problems (of marginal populations), the following appear to be most urgent:

"1. A great enlargement of the home demonstration service staff with a more definite assumption by home demonstration workers of the responsibility of rendering more aid to a larger percentage of the marginal families, together with any needed modifications of their procedures and programs to better fit the needs of the marginal group.

"2. The extension of more aid to subsistence farmers by farm agents and high school agricultural teachers, rather than confining their efforts so exclusively to commercial farmers, such aid to include efforts for participation by subsistence farmers in several needed types of cooperative enterprises." Garnett, W. E. Blacksburg: Virginia Polytechnic Institute Extension Service News. 18. No. 3. January 1936

[10] *Fourteenth Annual Report.* Chicago: National Committee on Boys and Girls Club Work. 1936. Pp. 32. Mimeo.

Joy, Barnard D. *Statistical Analysis of Trends in 4-H Club Work.* Washington, D.C.: U. S. Department of Agriculture. Extension Service Circular 247. August 1936

and experiments in this field are being carried on in many states. This movement has undoubtedly been very greatly stimulated by the increase of older youth in rural areas caused by the depression, and promises to be a significant phase of the extension program.

An important movement in rural adult education which has been incited by the depression is that for open forums and small discussion groups for considering national economic problems related to agriculture, international relations, etc. The forums have been held mostly in rural high schools, under the leadership of the agricultural teacher, the principal, or the county agricultural agent, while the smaller discussion groups have been held in various places usually under the direction of the county agent. These topics related themselves to the county program planning involved in the procedure of developing policies for the Agricultural Adjustment Administration. Thus broader and local problems were sometimes combined in community and county discussion groups, as in Virginia[11] where this movement was statewide. This movement has been systematically promoted through the extension services by the U. S. Department of Agriculture and in the schools by the U. S. Office of Education, with the cooperation of the state organizations. There is danger, of course, that such a movement might become merely the vehicle for political propaganda, but there seems to have been real effort to have both sides represented in the discussions.

For many years the teachers of vocational agriculture in the rural high schools held short courses for young farmers. Recently, this work of the rural high schools in adult education has been expanded into weekly classes of various sorts by many

[11] Hummel, B. L. *Summary of Group Discussion and County Agricultural Program Planning, State of Virginia, 1936.* Blacksburg: Virginia Polytechnic Institute. 1936. Pp. 15. Mimeo

schools which are experimenting with programs of adult education. This has been stimulated by the adult education movement and by assistance given through the WPA. Although this work is still in an experimental stage, it has progressed far enough to show the interest in it and the increased number of older youth furnishes a larger constituency than was formerly available.

The research projects which suggest themselves from the changes in extension work and adult education described above have to do with the possible social effects of these trends, and thus with the indirect rather than with the direct effects of the depression.

The experiments with discussion groups suggest the need of determining the best procedures in their organization and conduct; the best topics and those which should be avoided; the factual literature which should be supplied the participants and the best sources for such material; the rôle of the local and state libraries and of the extension service in furnishing such material. If suitable arrangements could be made for observing the organization and development of discussion groups, much might be learned as to the emergence of leadership through this method and of changes in attitudes which result from these discussions.

The systematic campaigns which have been conducted in the education of farmers with regard to their business and economic problems suggest the feasibility of attempting to develop knowledge tests on economic facts which might be applied before and after such campaigns. If these could be satisfactorily developed, they might be used to show the difference in the results obtained by different types of campaigns, such as those depending chiefly upon speaking by experts, newspaper articles, circular letters, news bulletins, discussion groups, or other methods.

The absorption of the time of the county agricultural agents with matters connected with the organization of the AAA and other governmental administrations raises the question as to the effect that this may have had on the personal opinions of the agents, with regard to the validity and value of the legislation which they have been required to help put into effect. It would be worthwhile to learn whether their original views with regard to various enactments have been changed by their participation in organizing their counties for their administration, and to what extent do their present views on these matters reflect the local public sentiment of their counties. Do those agents who are compelled, by virtue of their position, to promote policies which they personally believe are economically unsound or undesirable, feel that they can maintain their loyalty to the policies of the state colleges and the federal department of agriculture; or do they feel that if this situation continues they would prefer to get into some other line of work? These are questions which are of fundamental importance in maintaining the morale of so large a body of men who have such important positions of leadership, and in determining the future policies of the extension service, but are questions the answers to which could be obtained only through establishing thorough rapport with the agents and assuring them of anonymity in their replies.

Has the increased contact which the county extension agents have had with other county and federal administrators resulted in a broadening of their conception of their jobs, in broadening their programs of work, or in stimulating their cooperation in the development of councils or conferences of various executives and representative citizens for the consideration of local public welfare problems, and for social planning in their counties? Investigations of the extent to which this has occurred, under what circumstances, by what methods, and with what results, would be a distinct contribution to our knowledge of the effect which the depression has had indirectly on the integration of various agencies for rural betterment.

ON RURAL SOCIAL WELFARE WORK[12]

Prior to the World War there was little social welfare work in rural counties throughout the United States, the existing poor relief being conducted mostly on the basis of temporary assistance to keep families from starving or freezing. During the war the Home Service work of the American Red Cross was instituted for the families of enlisted men and introduced new methods and standards of social work into many rural communities. After the war it was in many cases extended to the civilian population, and this gave rise to a better organization of rural social work in many scattered counties. However, the promotion of better social work was abandoned by the National Red Cross as a general policy and left to the discretion of individual chapters as to its continuance. The industrial depression necessitated the organization of county relief administrations throughout the country and resulted in the introduction of methods of work and a general concept of social work which gave social welfare work a new position as an essential service of rural life.

There has been a large amount of research work done with regard to the relief situation in rural areas by the FERA and WPA and the state supervisors of rural research cooperating with these administrations. Most of this has, however, been concerned with an analysis of the relief population and its characteristics, and has given relatively little attention to the larger aspects of the effect of relief work and the place of public welfare organization in rural life. Unfortunately, there is no comprehensive publication showing the general status and organization of public welfare work in rural counties throughout the United States prior to the depression, which may serve as a base with which to contrast the present situation. Such records

---

[12] See also White, R. Clyde and Mary K. *Research Memorandum on the Social Aspects of Relief in the Depression;* and Chapin, F. Stuart and Queen, Stuart A. *Research Memorandum on Social Work in the Depression.* (monograph in this series)

exist in the reports of state departments of public welfare, in the reports of special commissions for individual states, and in publications of the U. S. Children's Bureau, but it would be highly desirable to have a careful description of the status of rural social work in 1930 made for each state as a background for studying the changes since then.

One effect of state and national aid in relief administration which has been noted by correspondents from several states is the decline of local responsibility for welfare work. Whereas, formerly, rural communities supported their own local welfare organizations or depended upon the town or county public welfare administrations for relief work, they have more and more shifted this responsibility to state and federal administrations. This change was stimulated by the assumption of county administration by representatives of the federal relief administration in some states. There has been a notable decline in the support of local agencies largely due to the sheer impossibility of their coping with the magnitude of the service necessary. This is a matter of common knowledge and has been revealed in numerous reports on individual counties made by the research workers of the federal and state relief administrations, but which have not been published. A thorough analysis of the present status of local private welfare agencies and of the feeling of local responsibility for public welfare work and of the relation of private agencies to it, would be helpful in determining future policies and programs.

Although this dependence on state and federal aid for financial support has possibly diminished the sense of local responsibility, it has brought about a recognition that public welfare administration has become a necessary part of rural government which must be administered more efficiently to conserve the cost to the taxpayer. It has come to rank with roads and education as a chief item in the cost of rural government, and as cost increases the taxpayer gradually demands better administration and recognizes the need of a better qualified personnel. In most of the United States the rural public welfare work is organized on a county administrative unit, and the county administration has been greatly strengthened in all states by the fact that it is the smallest unit with which federal and state relief administrations have dealt. However, in New England and several of the northeastern states the township unit has been the traditional agency for public welfare work, and in spite of reports of various state commissions recommending a change, the experience with relief administration during the depression does not seem to have created any general sentiment among farm people for abandoning it in favor of a centralized county administration.[13] Although professional social workers and students of rural government seem almost unanimous in favoring a change from a township to a county unit of administration, there seems to have been no careful study of whatever merits there may be to the objections which farm people make to such a change, or of how their opinions have been or may be altered so as to favor it. Here is a field of research which should be carried on by those who have the farmers' point of view and also appreciate what is necessary for efficient public welfare work.[14] If, by legislation, county administration is forced upon farm people who favor township administration, without a change of attitude on their part, it will meet with apathy and a lack of public support which will make its success difficult.

Another innovation in rural welfare work which has been

---

[13] Cf. *State and Local Welfare Organization in the State of New York.* Albany: Governor's Commission on Unemployment Relief. Legislative Document No. 56. 1936. P. 97; Matson, Opal V. *Local Relief to Dependents.* Michigan Commission of Inquiry Into County, Township, and School District Government. Detroit: Detroit Bureau of Governmental Research. September 1933

[14] For an outline of a research project which would be helpful in such a study see Brown, Josephine C. "County Organization for Rural Social Work." *Research in Rural Social Work.* New York: Social Science Research Council. Bulletin No. 5. July 1932. P. 94. (John D. Black, editor). This bulletin contains outlines for various research projects in the general field of rural social work.

introduced and fostered by the state and federal relief administrations has been the employment of professionally trained social workers. Prior to the depression the number of trained social workers employed in rural counties was relatively negligible for most of the United States. Hundreds of them were given positions as county relief administrators, and hundreds of local relief workers with no professional training were given scholarships in schools of social work to obtain as much as possible in one or two semesters or in special short courses. The reaction of rural people to the employment of trained social workers from outside the locality has been by no means uniform or entirely favorable, but will doubtless change as they become aware of the need for experience and training in this field—just as it has changed from a demand that local people be employed as school teachers and road engineers to a recognition of the advantage of appointing persons with the best qualifications for these positions wherever obtainable. A study of this matter is being conducted by the WPA, but it will be desirable to have studies of the problems involved made in several states and in different types of regions to obtain as objective a view as possible of the successes and failure of professionally trained social workers in rural communities and of the attitudes of the local people toward them. It would be expected that the very considerable increase in the number of these trained workers would have resulted in a more general recognition of the desirability of entrusting social work to them, and although this attitude is general among the state welfare administrations, it is by no means clear that it has the endorsement of the mass of the rural people. Research in this field is, therefore, desirable as a means for creating public opinion which will support a better trained personnel. A considerable mass of material pertinent to such research exists in the files of the federal and state relief administrations if access to it can be obtained. It would be fortunate if such research might be conducted and published by agencies independent of them and which would feel free to publish the facts as found without institutional bias or conforming to administrative policies.

One of the most outstanding effects of the unprecedented amount of public relief given in rural areas seems to be the changed attitude toward the acceptance of relief, which appears to have arisen in many rural communities. Formerly, to accept public relief was a social stigma and was considered the last step in the loss of self-respect. As relief became generally available this seems to have changed to an attitude of trying to obtain as much relief as possible because society owed the unfortunate a living. This was more common in what have been termed "stranded communities" where the ordinary means of employment had largely disappeared with no prospect of revival. In such areas even county relief administrators, who were expected to conserve the expenditure of relief, vie with each other in obtaining as much state and federal aid as possible as a means of placating their clients.

One correspondent who has been in close touch with the situation in an eastern state writes:

I believe there is little doubt but that the various efforts of the government to relieve the effects of the depression have resulted in inculcating into our population, both rural and urban, the psychology of dependency. A number of close observers with whom I have discussed this matter, state freely their belief that Federal aid in the depression, particularly direct relief, has been hurtful, in that many of those who received it have lost a disposition to do honest labor and bear their own responsibilities.

This is not the place to discuss the pros and cons of this much disputed matter, for in some cases it is true that persons on relief have sometimes been blamed for not being willing to work for starvation wages for temporary jobs rather than to accept the relative security of relief. It does, however, reveal a very general impression of the seriousness of the loss of morale which has probably taken place, and which varies widely in different communities and areas. It would be valuable to have a considerable number of case studies, of those receiving relief for some time, on the attitudes and character of individuals and of families, for it seems possible that social status may be much more seriously affected by being a public charge in a rural community where people are better known to all, than in a city. For similar reasons, it would be desirable to have studies of the differences in communities with regard to the acceptance of relief and the attitude of expecting and demanding public support, for as much difference in this probably exists between communities as between individuals, and a knowledge of the conditions and circumstances which increase or decrease these attitudes would be helpful in rebuilding community morale. The files of the local representatives of the Rehabilitation Division of the Resettlement Administration must contain many case records whose histories[15] might be followed up, as would the files of every county relief administration. State supervisors of rural research of the WPA, and state relief administration supervisors would be able to point out individual communities which differ widely in their reaction to relief.

One of the most important outcomes of the depression may have been a new conception, upon the part of rural welfare officials and rural people generally, of the objectives of giving relief and of family case work. To what extent have they come to see that, although many families who are perfectly capable of managing their own affairs have been forced to seek public aid during the depression, a larger proportion are families who have never been very successful, and that it is as important

---

[15] For an example of the characterization of families as to their possible rehabilitation see Burgess, P. S. and Tetreau, E. D. *Spot Survey of Sixty Families Referred for Rehabilitation to the Resettlement Administration by Pinal County Board of Public Welfare.* Tucson: University of Arizona Agricultural Experiment Station. Social Research Division of the Works Progress Administration. Rural Rehabilitation Division of the Resettlement Administration, and Pinal County Board of Public Welfare Cooperating. June 1936. Pp. 32. Mimeo. More complete case histories of itinerant agricultural laborers are given in *Migratory Labor in California.* San Francisco: State Relief Commission of California. Pp. 133-200. July 1936

to help them to become independent and self-reliant as to give them temporary aid? If the depression has shown rural people that it is more important to rehabilitate a family than to dole out groceries, fuel, and clothing for their temporary sustenance, it will have effected a radical change in their point of view which will undoubtedly lead to a different type of administration. It might be well worthwhile to study this problem by means of attitude tests in communities of different types and among different classes of rural people.

This whole topic leads to the question as to whether sufficient analysis has been made of the causes of poverty in rural areas. Poverty has been studied mainly in cities and under industrial conditions. Although there is a mass of data with regard to the causes of success and failure in farming as revealed by farm management studies, there has been little concerted effort to determine the outstanding causes of rural poverty in areas of different types through a consideration of both the economic and social factors and the individual differences of persons and of families, and the factors which have conditioned their incompetency or maladjustment. The studies of the relief population and its characteristics made by the FERA, the WPA, and their cooperating rural research supervisors, and the state relief administrations, have a wealth of data bearing upon this point, but these studies need integration and analysis to bring out more clearly the primary factors responsible for rural poverty under varying conditions. One experiment station sociologist has already outlined a study of this sort and is organizing a committee of his state planning board for a study of the problem. This suggests that state planning boards might well give attention to such fundamental problems of the human resources and liabilities of their states,[16] for, after all, it is the

---

[16] As good an example of this sort of comprehensive social and economic analysis as is available is *Economic and Social Problems and Conditions of the Southern Appalachians*. Washington, D.C.: United States Department of Agriculture. Miscellaneous Publication No. 205. January 1935. Another example of

welfare of the mass of the people for which all analysis of state resources is made. If there are fundamental weaknesses in the social structure which make poverty inevitable, it would be well to reveal their causes as a basis for sound planning of any sort.

SUMMARY

Although the county public health services were temporarily contracted, the evidence seems to show that rural services in general were considerably expanded as a result of the depression, and that they will probably become of more importance in rural life than formerly. The topics suggested for research are quite diverse. In the field of health they include the investigation of the number of rural physicians and the need for more county public health units. Concerning extension work, it will be worthwhile to study the methods of starting and conducting discussion groups as a method of adult education and their possibilities for the development of rural leadership, and to determine the extent to which the work with other agencies during the depression has widened the concept of extension work. Social welfare work has had an unprecedented expansion and gives rise to several problems: What are the merits of township public welfare administration, what are the attitudes of rural people toward it, and how may they be changed? To what extent have professional social workers been accepted by rural people and what are their attitudes toward them? Has there been any change in the opinion of rural people as to the desirability of constructive case work with disadvantaged families as a phase of relief work? Finally, there is need of a more exhaustive study of the causes and means of alleviating rural poverty, considering not only the economic factors but also those of individual differences and social conditioning.

a study of the human and other resources of a state made under an unofficial commission is *Rural Vermont: A Program for the Future.* Burlington: Vermont Commission on Country Life. 1931

Chapter VIII

Attitudes toward the Future of Agriculture

IF THE depression has had any effect upon the attitudes of farmers toward the future of agriculture as a vocation and as a mode of life, this might prove to be one of its most important consequences to rural society. Obviously this will vary widely according to how severely the depression has affected farmers; from the northeastern states where relatively few farm operators were compelled to apply for relief, to the far western drought areas where many were forced to migrate. Individuals familiar with local conditions in various states believe that the better class of farmers were adjusting themselves to conditions and felt confidence in the future of agriculture, whereas the poorer farmers were increasingly discouraged and radical. In attempting to understand their attitudes we must recognize the fact that the agricultural depression started in 1921 and that a gradual recovery was under way when the industrial depression of 1930 plunged the farmers into deeper trouble. It is difficult to obtain data for a satisfactory answer to our question. Even the actual behavior of farmers may not reveal their motives or attitudes. Thus the fact that there is lively bidding for leases on Iowa farms may mean that tenants have faith in farming, or it may mean that they feel they do not know how to do anything else and that there is no chance for employment in cities.

Several topics are significant in attempting an evaluation of the attitude of farmers toward the future of their vocation. Has their standard of living been permanently lowered, or are they hopeful of maintaining it? Do they feel that they cannot succeed except with increased aid of the government? Do memberships in farmers' cooperative associations and in distinctive farmers' organizations reveal losses or gains which may indicate their faith in their business? Has their position during the depression as compared with unemployed wage earners given them a new sense of security? Evidence on these points may indicate the trend of their vocational attitudes, and will now be considered.

EFFECT ON CHANGES IN THE STANDARD OF LIVING[1]

To what extent have the standards of living of rural people been lowered by the depression and will this affect them permanently? Only a direct comparison of the standard of living during the depression with that prior to it will satisfactorily answer this question. Fortunately Kirkpatrick[2] has been able to do this for 313 families in three Wisconsin counties. In his summary he states:

The prosperous farmers of 1929 are down to marginal standards of family and community living today; the marginal at that time are submarginal now, and the submarginal then are eking out a bare existence with aid from public sources.

Furthermore the data indicate that there are accepted optimum standards of living which are being tenaciously held by many farm families. In some instances the accepted standards are being maintained by raising more of the living from the farm, thus making the available cash go farther. In other instances relatively more satisfactions are being obtained from the so-called intangible aspects of family and community living.

[1] For bibliography see Williams, Faith M. and Zimmerman, Carle C. *Studies of Family Living in the United States and Other Countries*. Washington, D.C.: U. S. Department of Agriculture. Miscellaneous Publication No. 223. 1935. P. 617

[2] Kirkpatrick, E. L., Tough, Rosalind, and Cowles, May L. *How Farm Families Meet the Emergency*. Madison: University of Wisconsin Agricultural Experiment Station. Research Bulletin 126. January 1935

We have commented on the last observation in considering recreation. It would be very desirable to have similar surveys made of the standard of living of families which had been studied prior to 1930 wherever possible. Several studies of the standard of living on farms are now being made under the direction of the Division of Farm Population and Rural Life of the U. S. Bureau of Agricultural Economics, and to the extent that they are made in counties whose economic and social conditions are comparable with areas in which standard of living studies were made prior to 1930, it may be possible to obtain an approximation of the decline of the standard of living in those sections and what changes have occurred in it. It is evident that we shall not be able to answer questions with regard to changes in the standard of living until data upon this topic are collected in well selected sample areas, either annually or at regular intervals of three to five years, so that there may be a basis of comparison in times of depression or prosperity.

The better prices for farm products during and just after the World War resulted in an improved standard of living which was also incited by the ownership of more automobiles and more contacts with town and city. The sudden slump in farm income in the twenties with an increased appetite for more of the material comforts of life undoubtedly produced a certain sense of frustration which was climaxed by the industrial depression. The disparity between psychological standards or wants and the levels of economic consumption possible has undoubtedly increased on account of the urban influence on rural attitudes.

It has been generally observed that the industrial depression has resulted in an increased dependence upon the products of the farm. There has been a general increase in canning and the preservation of foods on the farm. In some sections there has been a notable increase in home gardens, fostered by both the extension services and relief agencies. However, when one remembers the enthusiasm for canning and gardening during the World War and their subsequent gradual decline, one cannot be sure as to whether this will have any permanent effect upon the standard of living of the poorer class of farm families who are most benefited by producing more of their subsistence.

Although his data give no indication of the effect of the depression, one of the most important findings of Dr. Woofter's[3] study of the tenant on the cotton plantation is that the net income of the tenant family increases with the amount received as income from home use production, and that this is inversely related to the proportion of the acreage in cotton (the income from products used at home decreasing with the proportion of acreage in cotton). It would seem desirable to make similar studies of the part which raising products for home use forms in the total income of tenant families, under various conditions in each state, as a means of demonstrating the desirability of devoting more acreage to production for home use and thus developing a better standard of living.

There has been a notable decrease in the number of farm telephones in many states and increased use of the driving horse in place of the automobile. Automobiles were disposed of or unused because of the cost of repairs and gasoline. This transition back to the horse and buggy was not so easy in these days of hard roads designed for motor traffic. As a result the people affected have probably attended social institutions less and there has probably been an increase of social isolation.

One who is well acquainted with conditions in Oklahoma states that improved lighting systems using electric or gas generators have been discarded and the old oil lamps have been rescued from the attic and restored to use. On the other hand, in New York, Iowa, and other states there has been a considerable

---

[3] Woofter, T. J., Jr. *Landlord and Tenant on the Cotton Plantation*. Washington, D.C.: Works Progress Administration. Division of Social Research. Research Monograph 5. 1936. See Appendix, Table 40 and p. 84

increase in rural electrification as a result of federal aid. The agitation for rural electrification and stimulus being given by the Rural Electrification Administration and the Tennessee Valley Authority will doubtless result in a very considerable increased use of electricity on the farm and in farm homes.

Instead of reducing the hours of work, as has been the result for city wage earners, the industrial depression probably has caused the farmer to work longer hours, and probably has compelled farm women and children to do more farm work. It is important to determine in detail the amount of farm labor by women and children as a factor in the family income, as to relative income of agriculture as compared with other industries, and as a factor limiting the standard of living.

In certain sections of the South there is evidence that, because of the type of relief administered, such as medical services and dietary instruction, many of the underprivileged share croppers in more backward communities have gained new ideas of what is an adequate standard of living, and that their standards have been raised. A study of what changes have occurred in the standard of living of rehabilitation clients might furnish evidence on this point.

Several observers have noted a decrease in thrift and a tendency to enjoy modern luxuries, regardless of dwindling resources or insecurity of income, but whether this applies to most of the people or only to the examples observed is an open question. A consideration of the available data makes it appropriate to raise the question of whether these effects will endure well after the depression. Or with better income will not rural people resume their former plane of living and seek to improve it? The latter effect seems the more probable. It is to be remembered, however, that since 1921 farm families were not able to enjoy the fruits of urban industry to the same extent as urban families, and that during the twenties, when urban homes were increasing their comforts very materially, most farm families were unable to do so. It would seem that this must have had the effect of increasing the lag between rural and urban progress, so far as material comforts are concerned, and thus creating greater discontent among farm people.

On the other hand, it must be recognized that rural people are now able to enjoy many comforts and conveniences not available until recent years and that with larger volume and lowered costs of production these will be more generally attainable in the future. Electricity in the home and barn, central heating plants, running water and indoor toilets, iceless refrigerators, and many other conveniences in the home, to say nothing of better communication by means of automobile, telephone, and radio, make possible a standard of living which was not dreamed of by the wealthiest farmers a generation or two ago. It is probably possible to maintain as good or better a standard of living on the farm, as in the city, so far as material goods are concerned, so that the relative satisfaction in farm life will increasingly depend upon what economists call the "psychic income," the values received from the non-material goods of life which are chiefly conditioned by education and cultural environment. Despite the fact that money income and the possibility of attaining wealth in farming will be relatively less than in other vocations requiring equal ability, it is likely that the values of the non-material goods of farm life will be relatively more important, and it is possible that in the future there will be a gradual increase in the proportion of men who enter farming because they prefer it as a vocation, rather than because they inherit the farm or do not know how to do anything else, as in the past. This being the case, it is important that attention be given to maintaining the facilities for a satisfactory cultural environment, schools, churches, libraries, organized social life, medical service, etc., and that any effects of the depression on these institutions be clearly recognized so that necessary steps may be taken for their maintenance.

INCREASED DEPENDENCE ON GOVERNMENT

From many well-informed correspondents comes the opinion that the depression has produced an increased dependence on governmental agencies and on legislation for improving rural conditions, and that farmers are showing less initiative to work out their own salvation through cooperative effort. The effect of federal aid for relief and of the Agricultural Adjustment Administration's program has undoubtedly been in this direction. In one state the farmers were at first doubtful with regard to the propriety of receiving AAA payments, but soon they were glad to accept them, and their discontinuance upon the suspension of the act produced a very definite resentment. This change seems to have been very general in the states most affected, as evidenced by the willingness of the Congress to enact new legislation to continue benefit payments. There is some evidence that this attitude is particularly prevalent among the poorer and less educated farmers and in states where they are numerous. One southern observer notes the tendency for this class to vote for "panacea promisors and demogogues." Even in the eastern states, where the effect of the depression was not so severe, one who is in close contact with thousands of farmers feels that there has been a tendency to accept any plan which promised immediate improvement of income rather than to consider measures which would be of most permanent and long time value, or a tendency to live for today rather than to plan for the future. On the other hand, there is a very general feeling that the activities of federal and state governments during the depression have decidedly strengthened the feeling of farmers that they can improve their condition only through collective action.

There is also a very general impression that whether the government's efforts to aid agriculture have been wise or not, they have furnished a convincing example of the possibility of the control of agricultural production. One correspondent writes:

Farmers have waked up to realize that industry has been curtailing production in order to keep prices up, even though it has resulted in the displacement of labor which had to be cared for by the federal government or other community agencies. Perhaps now the farmer has come to see that he can curtail production to bolster prices and let the federal government take care of his displaced labor force on a comparable basis with industry.

Undoubtedly there has been a trend in this direction and its total effect in changing the philosophy of the farmer with regard to his economic relations may have far reaching consequences and warrants careful study.

Just how to obtain any accurate measure of what effect the depression has had on these attitudes is a difficult problem. A close analysis of the votes of farmers upon the program and policies of the AAA, by comparing counties with different types of agriculture and differently affected by the depression and by drought, might reveal significant differences. A study of the editorial attitude of leading agricultural papers and of their comment on governmental programs and policies, would give a reflection of the differences in public sentiment in different areas and might be revealing. A carefully prepared attitude test administered to large numbers of farmers of all sorts in well selected sample areas, would probably be the most satisfactory method of determining their present attitudes; but to what extent they have changed as a result of the depression would be largely a matter of inference because of no similar data previously obtained. In spite of the expense involved it would seem well worth while to undertake a rather extensive study of the attitudes of farmers not only on this question of dependence on government, but of their attitudes toward farming as a vocation and mode of life, their sense of security in it, etc.

GROWTH OF FARMERS' COOPERATIVE ASSOCIATIONS

If farmers are placing more trust in the government as the agency for solving their economic problems, it would seem possible that this might affect the growth of their cooperative associations, more particularly those for marketing. The best evidence on this subject are the statistics published by the Cooperative Division of the Farm Credit Administration.[4] The total number of farmers' marketing associations shows a decline from 1932 to 1935, but a study of the trend for associations dealing with different commodities shows that this decline was chiefly among the local associations marketing grain and livestock. On the other hand, the terminal agencies handling grain and live stock cooperatively showed no appreciable decrease in the volume of business, other than might be accounted for by less production. The cooperative associations marketing practically all other commodities showed definite growth in the volume of business handled during the depression, with occasional decreases in 1932 or 1933, but there is no evidence of a definite decline during the industrial depression.

The largest growth in the number of cooperative marketing associations was from 1915 to 1920. The fact that their number remained fairly constant from 1920 to 1930 during the agricultural depression of those years shows that they had become firmly established.

On the other hand cooperative associations for the purchase of farm supplies have shown a steady growth since 1925, and a larger growth from 1932 to 1935 than in the previous years.

Evidence tending to show that the loss of support of farmers' cooperative associations as a result of the depression appears to be overbalanced by considerable evidence that it has given their development a healthy stimulus. It should be observed, however, that the members of cooperative associations are mostly among the better type of farmers and it would be important to make studies of individual associations and to determine whether there

---

[4] Elsworth, R. H. *Statistics of Farmers' Cooperative Business Organizations.* Washington, D.C.: Farm Credit Administration. Cooperative Division. Bulletin No. 6. May 1936

had been any gain or loss of particular types of farmers according to economic status.

It will also be desirable to have studies of the growth or decline of cooperative associations for the past five years in individual states, for it appears that in some states there is a definite opinion that cooperation has lost ground, while in others it is believed to have gained. The causes of this, whether they be due to differences in products, in organization and management, or in the types of farmers, should throw light on the problems of cooperative associations.

It is a matter of common knowledge that many of the leaders of the cooperative movement have been opposed to the policies of the federal administration with regard to the control of production and any measures which affect the control of prices. This opposition is based on the ground that governmental interference will produce only an artificial situation and that any permanent advancement in obtaining a larger share of the national income for agriculture must come from the better organization of the industry through farmer controlled cooperative associations.

MEMBERSHIP IN FARMERS' ORGANIZATIONS

Another measure of farmers' confidence in their business may be had in the growth or decline of membership in strictly farmers' organizations such as the Grange and the Farm Bureau. Neither of these organizations publish their membership, but both the National Grange and the American Farm Bureau Federation publish receipts from states for membership fees, from which the number of members can be rather accurately determined. Table III, page 133, gives the estimated membership for both organizations since 1920, when the Farm Bureau Federation was founded. There is reason to believe that the figures for Farm Bureau membership do not represent the total in county farm bureaus, for Burritt[5] states that in 1920 there were 826,816 members in 37 states and that in 1921 there were nearly 967,279 members in 46 states.

The Grange showed a slight decline in 1933, from which it has more than recovered in the last two years, but, in general, it showed no material change in its membership during the 16 years. However, the changes in membership in the Grange differed widely in various regions. Thus Washington, Oregon,

TABLE III

ESTIMATED NUMBER OF MEMBERS OF GRANGES AND OF FARM
BUREAUS: UNITED STATES, 1920-1935

YEAR	GRANGES[a]	FARM BUREAUS[b]	YEAR	GRANGES[a]	FARM BUREAUS[b]
1920	541,158[c]	317,109	1928	585,253	301,699
1921	587,091[c]	466,423	1929	576,563	301,733
1922	663,605	363,483	1930	601,166	321,196
1923	601,087	392,583	1931	563,877	276,053
1924	647,056	301,747	1932	597,831	205,350
1925	584,943	314,476	1933	575,068	163,247
1926	581,667	278,760	1934	583,005	222,558
1927	583,912	272,049	1935	604,390	280,917

[a] From *Journal of Proceedings of the National Grange.* Except 1920, 1921
[b] From statement of receipts for membership furnished by the Secretary of the American Farm Bureau Federation.
[c] From Wiest, Edward. *Agricultural Organization in the United States.* P. 375

Idaho, and California showed strong gains during the industrial depression, whereas some of the eastern states, where the Grange is strongest, e.g., Pennsylvania, declined in membership.

In contrast, the Farm Bureau, after the initial boom of the first two years of the national federation, declined steadily to 1927, then gained until 1930, declined to the lowest point in 1933, and in the last two years has increased to about the same number as in 1927 and 1931.

As far as the receipts of the American Farm Bureau Federa-

[5] Burritt, M. C. *The County Agent and the Farm Bureau.* New York: Harcourt, Brace & Co. 1922. Pp. 235, 245

tion show there was little difference in the decline or growth between the four regions which it recognizes, northeast, central, southern, and western.

Wiest[6] gives a table showing the membership of the Grange from 1874 to 1922. After the depression of 1893 the membership of the Grange had a sudden drop for the next three years and then grew gradually up to 1906, had a large increase in 1907, and after the depression of that year dropped for the next two years. From 1911 to 1921 there were some ups and downs, but no definite trend, until 1921 when there was a considerable increase (8 per cent) and in 1922 there was a gain of 15 per cent over 1921. The increase in the last two years corresponded with the sudden growth of the Farm Bureau and was doubtless the result of dissatisfaction arising from the decline of prices in those years.

Sufficient evidence has been given to show that the growth and decline of these farmers' organizations need careful analysis by states and regions, before we can understand the significance of the changes in membership occurring during the depression. This is particularly true of the Farm Bureau which in some states is a radically different type of organization from what it is in others.

From the evidence cited, it would seem probable that, as a result of the reaction from the depression, there will be a considerable growth of these organizations in the next few years. If so, it will be important to determine to what extent this is built up by playing upon the class feelings of farmers aroused by their experiences during the depression. In fact it is high time that as thorough study be given to the recent history and

---

[6] Wiest, Edward. *Agricultural Organization in the United States*. Lexington: University of Kentucky. April 1923. P. 395. For a careful analysis of the growth and decline of farmers' organizations see McCormick, Thomas C. *Rural Unrest: A Sociological Investigation of the Rural Movement in the United States*. University of Chicago. Unpublished. Ph.D. dissertation. 1929

policies of the leading farmers' organizations as has been given the trade union movement. An interesting comparison might be made of the trend of trade unionism—from the Knights of Labor, "idealistic, humanitarian, and political,"[7] during the last two decades of the nineteenth century, to the craft unions of the American Federation of Labor characteristic of this century, to the present trend toward industrial unionism—with the development of farmers' organizations, the Grange and the Farm Bureau—as contrasted with the growth of the cooperative associations.

In interpreting the membership of these organizations it must be remembered that they include the better class of farm families, and that, in general, they do not reach any large proportion of the poorer farmers. One of the most important factors affecting the membership and policies of the Farm Bureau which arose out of the depression was the rise of such direct action movements as the Farmers' Holiday Association, which have undoubtedly been influenced by the philosophy and techniques of trade unions. In an able paper on this topic Dr. Carl C. Taylor makes the following observations:

Finally, direct action movements, being made up of left-wing groups, have set themselves off strongly against right-wing or middle-of-the-road farm organizations and have thus done material damage to these more permanent organizations. It should be noted here that the effects of a left-wing organization on the established class organization have been in both directions; such organizations have at times driven the established organizations further to the right and at times pulled them further to the left. In every instance it has jeopardized the strength of the established organization for a period of time.[8]

---

[7] *Cf.* Hoxie, R. F. *Trade Unionism in the United States.* D. Appleton and Co. 1917. P. 87

[8] Taylor, Carl C. "Notes on Some Theoretical Aspects of the Effect of Direct Action Farmers' Movements on Farmers' Organizations." *Social Forces.* 12:386-387. March 1933. This is but an abstract of a paper presented by Dr. Taylor before the Rural Sociology Section of the American Sociological Society. Summer Conference, Chicago, June 29 1933. It is hoped that the complete article may be published shortly, as Dr. Taylor has been the best student of these phenomena for many years.

The effect of these more radical movements has been felt even in those states where the established farmers' organizations are strongest and their influence on the future policies of these organizations will be among the most significant results of the depression. It is highly desirable, therefore, that the history of the movements be recorded and interpreted while the sources of information are available, for the most important data are unpublished and must be gathered from individuals. It is particularly important to attempt some measure of the actual membership in these movements, the type of farmers and their location, for the facts are usually grossly exaggerated in the press.

THE FARMER'S FEELING OF SECURITY

Much has been said with regard to the larger degree of security which farmers have during a depression, as compared with wage earners, and this has been particularly evident to city wage earners who have come from farms. Has the farmer shared in this feeling and is he more or less satisfied with his position as compared with that of city wage earners? The opinion has been advanced that the average or better class farmer has an enhanced appreciation of the advantages of farming as a vocation and has hope for the future. In the Corn Belt, in particular, there is supposed to have been an increased interest in the farm as a place to live and in better living on the farm. The lack of economic income to increase their standard of living in pace with that in the cities during the twenties, forced farmers to give consideration to putting more of their income into the home. Moreover, the agricultural extension services in several states made definite campaigns to encourage farm men and women to consider the problems of farm management in terms of obtaining a certain farm income as a means of improving the standard of living in the home rather than for expanding the farm business. This may have had an influence in turning attention to the advantages of the relative self-sufficiency of the farm during the industrial depression. One correspondent in the South writes:

> Another attitude which has impressed me somewhat is one which might be termed satisfaction with farm life. Many rural people have pointed out to me the predicament of their friends who had left the farming occupation to follow 'public works.' Many farm people seem to get considerable satisfaction out of the fact that they were not lured away during 'boom times' by the enticements of a regular weekly or monthly pay check. They feel that they are the best off in the long run.

On the other hand there is equal evidence that the class of farmers who were forced off their farms by the depression, either by being reduced from owners to tenants or from tenants to laborers, tended to become more radical and dissatisfied with the possibilities of agriculture. This is but natural and has probably been a dynamic of rural unrest after every depression. Thus a correspondent from Minnesota writes:

> With respect to the general attitude of the farmers I think it is safe to say that the substantial farmer, who was able to carry on in spite of falling prices and managed to adjust himself economically, very little change took place, whereas those who were less able to make the adjustment have apparently swung to the extreme left and intensified their hostility to urban people and urban institutions.

This possible tendency for the better farmers to maintain their position while the poorer farmers decline, may partially account for the fact that the increase in the number of farms has been largest, in small and in large farms, while medium sized farms have not increased so rapidly and have declined in their proportion of the total. However, the mechanization of farming and the number of small farms around cities are undoubtedly the chief factors in this change. This tendency to increase the proportion of large (500 acres or over) and small (under 20 acres) farms has been going on during the present century, as may be seen from Table IV, page 138. These changes in size of farms should be analyzed for each state and region, and will be useful for comparison with the figures of

TABLE IV
PERCENTAGE DISTRIBUTION OF FARMS BY SIZE (1890–1935), AND PERCENTAGE
INCREASE IN NUMBERS (1930 TO 1935): UNITED STATES[a]

SIZE GROUP	PERCENTAGE DISTRIBUTION						PERCENTAGE INCREASE
	1935	1930	1920	1910	1900	1890	1930–1935
Total	100	100	100	100	100	100	8.0
Under 20 acres	18.3	14.6	12.4	13.2	11.7	9.1	53.3
Under 3 acres	0.5	0.7	0.3	0.3	0.7	—	20.9[b]
3 to 9 acres	7.8	5.0	4.2	5.0	3.9	—	70.0
10 to 19 acres	10.0	8.9	7.9	7.9	7.1	5.8	22.0
20 to 49 acres	21.1	22.9	23.3	22.2	21.9	19.8	0.0
50 to 99 acres	21.2	21.9	22.9	22.6	23.8	24.6	5.0
100 to 499 acres	35.5	36.8	38.1	39.2	39.9	44.0	4.4
100 to 174 acres	20.6	21.4	22.5	23.8	24.8	—	4.6
175 to 259 acres	7.9	8.3	8.2	8.4	8.5	—	3.8
260 to 499 acres	6.9	7.2	7.4	7.0	6.6	—	4.8
500 acres and over . . .	3.8	3.8	3.3	2.8	2.6	2.5	6.5
500 to 999 acres	2.5	2.5	2.3	2.0	1.8	1.8	4.8
1,000 acres and over . . .	1.3	1.3	1.0	0.8	0.8	0.7	10.0

[a] Per cent distribution 1890–1930 from Abstract of the Fifteenth Census of the United States Table 16, p. 526; 1935 percentages computed from United States Census of Agriculture: Farms By Size. 1935. Press release. October 14, 1936
[b] Decrease

change in tenure, for there are striking differences between areas in the changes of size between 1930 and 1935. Thus in the North Central States the farms of 20 to 49 acres, which formed only 11.7 per cent of the total, increased 21.1 per cent in number, whereas in the South Central States, excluding Kentucky and Tennessee, farms of this size comprised 29.8 per cent of the total and decreased 11.1 per cent in number.

After a careful survey of the trend of agriculture in this country, Dr. T. C. McCormick came to the following conclusions in 1931:

A majority of the present farmers and their descendants, by severe competition, will be gradually forced out of the ownership of farm land. Most of them will move into the towns and cities; the rest will be reduced to directed tenants and farm laborers, who will become increasingly identified with urban laborers in (part time factory) employment, organization, wages, and manner of living.

The relatively small and selected group of surviving farmer-proprietors, compared with the mass of farmers in the past, will be the operators of large-sized business units, of more machinery and mechanical power, of greater capital investments, and of increased number of laborers. They will be more skillful financiers, organized into specialized subgroups, each a rather efficient guardian of its vocational interests. These farmer entrepreneurs, occasionally heads of agricultural corporations, will have higher incomes and greater wealth; will occupy more pretentious homes on highways convenient to towns; will have smaller and less stable families; and will be members of various special-interest social groups drawn from larger areas and centering chiefly in towns. . . . Because of a variety of cultural stimulations and higher education, such farm people will have widened interests and a more progressive and scientific outlook. . . . In a word, the farmer-proprietor class will approach in type and culture the business and professional groups of the towns and smaller cities.[9]

Whether we agree with this prediction or not it undoubtedly reflects the trend of farm operation during the present century, and gives cause to the apprehension of the small owner and tenant that his security is decreasing. It is quite possible that the depression has accelerated this tendency toward the control of the land by farmer-proprietors and its operation by directed tenants and laborers. Over against this trend, however, should be balanced the increased volume of loans of the Farm Credit Administration, the scope of credit which it offers, and the liberal terms of repayment; all of which make it much easier for a responsible farmer to finance his affairs without fear of dispossession during temporary adversity.

Ability to climb the "agricultural ladder" from the position of hired laborer, through tenancy, to ownership, has been an American tradition. To maintain this opportunity for vertical mobility is one of the most fundamental issues in our national policy concerning agriculture, and is essential to a free democracy. Certainly the examples which we have of a large scale,

---

[9] McCormick, Thomas C. "Major Trends in Rural Life in the United States." *American Journal of Sociology.* 36:733-734. March 1931

highly commercialized agriculture, approximating factory methods of production as, for instance, in California, do not seem as conducive to the general welfare as a more diffused control by individual operators.

It is, therefore, important that accurate information should be gathered in every region, state, and agricultural section, concerning not only the basic facts of tenure and relative success or failure in farm mangement, but also the attitudes of farm men and women and boys and girls toward agriculture as a vocation and as a mode of life. It is quite possible that families who prefer farm life, and who may be enabled to maintain a satisfactory standard of living, even though it affords less cash income, may be able to compete successfully with a more commercialized type of agriculture, if they are efficiently organized for collective action.

As already indicated, probably the best method of ascertaining these attitudes toward farming would be through an extensive survey conducted by means of attitude tests. It would seem particularly desirable to make frequent studies of the attitudes of rural boys and girls toward farm life, as they leave the grade school and as they leave high school. It should not be difficult to arrange with school administrations so that such tests could be conducted regularly in the same communities over a series of years.

Chapter IX

Effects on Distinctive Rural Attitudes and Rural Culture

EQUALLY important with the attitudes toward agriculture are those characteristic rural attitudes which distinguish rural from urban culture. Have these been affected by the depression; has the depression accelerated the process of urbanization, or has it tended to preserve and magnify distinctive rural values?

RURAL-URBAN RELATIONSHIPS

Traditionally, farmers have had a distinct antipathy to city life, yet in recent years there has been a steady urbanization of rural life. Although the depression undoubtedly checked the amount of patronage of cities by rural people, this is probably only temporary. On the other hand, there has undoubtedly been a growing realization of the interdependence of agriculture and the business and industry of cities. This is partly due to the larger number of farmers who are employed part time in towns and cities, and to the return of the city unemployed to the country, but it is also due to the fact that the breakdown of the farmers' foreign markets has made them feel their dependence on the markets of our own cities. Moreover, because of increased contacts of country people with city life, through more frequent visits to cities and through reading city newspapers and magazines, they feel more at home in the city and are more susceptible to the influence of its standards and values.

In view of the evidence that all rural young people cannot be satisfactorily maintained in the country and that cities are dependent upon rural immigration to maintain their population, it would seem probable that a change of attitude toward rural-urban migration of young people may have taken place. If this be true, it should be reflected in a changed attitude toward the curriculums of rural high schools and their programs of vocational guidance, which seems to be the case.

Incidental to this change in rural-urban relationships there has undoubtedly been a change in the farmers' political views. Although there has been a growth in the alignment of farmers and city wage earners in the farmer-labor party of two or three West North Central States, farmers have in general resented the payment to city workers of the relatively high wages which have increased their operating costs. The closing of foreign markets to agricultural products has shown American farmers that agricultural exports can be paid for only through imports of foreign manufactured goods. This being true, if American manufacturers are to be protected by tariffs—which enable them to maintain wages for their workers and thus expand home markets for farm products, with the result of higher costs for the goods purchased by farmers—the farmers very naturally reason that they should have some equivalent form of added compensation for their products equivalent to the protection afforded city industries. This was the reasoning behind the vetoed McNary-Haugen Act, of the proposed export-debenture plan, and of the compensation given to the Agricultural Adjustment Administration. Traditionally, the farmers of the northern states have been the backbone of the Republican party, but the last two national elections seem to show that this allegiance has been effectively broken. To what extent this has been due to a change of their views with regard to the advantages of a high protective tariff it is impossible to demonstrate, because of the confusion of issues involved in national political campaigns. It would seem, however, that it would be well worthwhile to make a thorough study of this subject by means of knowledge tests concerning the farmers' information about the tariff and its effects and with tests of their attitudes concerning it. Whether such a survey could be properly carried on by governmentally supported agencies is doubtful, but it would be entirely feasible under other auspices to determine rather definitely their present knowledge and attitude on this subject and it would be of fundamental importance.

ON CHANGED ATTITUDES TOWARD GOVERNMENT

We have already discussed the farmers' tendency to seek government aid during the depression, but their change in attitude toward government goes much deeper than that. Traditionally, farmers have felt that the less government the better and that any increase in the functions of government meant an equal increase in taxes, direct or indirect. The slump in their foreign markets has forced them to a new understanding and attitude toward international relations, and they have realized that only through the federal government can these conditions be ameliorated. Furthermore they have seen, as never before, that their interests are bound up with the whole national economy and that only through the federal government can many of our fundamental economic difficulties be satisfactorily regulated. The loss of ownership of thousands upon thousands of farms has seriously shaken their sense of security, and they are willing to support the federal government in measures seeming to promise larger economic security which they would have decried in the past.

The whole movement toward a more planned economy has, therefore, been met with a receptive attitude on their part, particularly with regard to those phases which directly affect agriculture, whereas formerly they would have viewed any such tendencies with suspicion. The planning of county programs of agriculture which has been carried on very extensively by the agricultural extension services in recent years, and the determination of county quotas under the Agricultural Adjustment Administration, have undoubtedly tended to develop an interest in economic planning and created a mind-set favorable to it, even though farmers may not have been clearly conscious of their influence. It is fairly obvious to them, whatever their theoretical allegiance to state's rights may be, that only through the federal government may legislation or planning be effective to meet many of the most fundamental economic problems which transcend state lines.

The shift in attitudes which have been conducive to a larger dependence upon government has been well expressed by Dr. Carl C. Taylor, chief of the Division of Farm Population and Rural Life of the U. S. Bureau of Agricultural Economics and recently assistant administrator of the Resettlement Administration, who has had a wide and intimate acquaintance with farmer sentiment for many years:

> I believe that the two opposite poles of attitudes developed during the depression among rural people are: on the one hand, programs such as the Agricultural Adjustment Administration, Soil Erosion Service, and Resettlement Administration have induced in rural people a habit of *calculation* as versus *hit and miss*. These agencies have taught planning, calculation and a knowledge of the factors which naturally fruit in certain economic products. On the other hand, hundreds of thousands of farm families cannot become habituated to relief, become transient migrants, not knowing where they are to be next year or what they are to do, work on various projects the objectives of which they do not know, and in general become as unstable as they now are, without at the same time developing attitudes of defeatism and maybe of unconcern which we have never thought of being typical attitudes among the rural people.[1]

It is significant that the attitudes attributed to both of these classes make them predisposed to a larger functioning of the government to meet their problems.

---

[1] In a personal letter to the author.

ON COLLECTIVE ACTION VERSUS INDIVIDUALISM

These same tendencies have broken down the former "rugged individualism" of farmers to a very considerable extent, and they are now much more united in a general conviction that their problems can be met only through collective action, although in many cases they are still far apart as to how this should be effected. In times of crisis the individual presumably turns to some superior authority, either that of an individual leader or the collective authority of an organized group, whereas in better times he cherishes his independence and magnifies the values of individualism.

This tendency is seen most clearly in the present situation with regard to the marketing of milk in the New York City milk shed. Having tried state control and found that it does not solve their problems without interstate agreement or federal control of interstate shipments, the milk producers of the state, both those who belong to rival milk marketing organizations and those who belong to none, now seem to be convinced that there is no solution of their problem without united action which will control the situation. At the time this is being written, the milk producers are again assembling a conference of all interests concerned to attempt to get together in a practicable procedure. The point is that the situation has forced them to the conviction that integration is essential and that there is no hope for the individual producer except through organization. Such changes in attitudes come only through long and costly experience.

The steady growth of farmers' cooperative associations throughout the whole period of the agricultural depression is also evidence of the gradual but increasing change in the attitude of the farmer from the individualism of the pioneer and of the farmer of prewar days (who could succeed through his own initiative) to a belief that only through organization and united action can he maintain his interests.

ON STRENGTHENING RURAL VALUES AND A DISTINCTIVE RURAL CULTURE[2]

The above discussion has dealt chiefly with changes in farmers' attitudes toward economic and political problems. The question might well be raised as to what changes, if any, have occurred in his appreciation of the peculiar values of rural life —of the immaterial culture of rural life which distinguishes it from that of the town and city.

On the one hand, there are certain changes, accentuated by the depression, which would seem inimical to the accepted rural values of former days. Thus the rapid increase of part time and subsistence farming near cities and the movement of city workers into country villages must have had a definite, if yet undescribed, effect upon the ways of life and ideals of the communities involved. The former ideals of thrift and the distinction of the indefatigable worker described by Williams[3] as characteristic of the Central New York farmer of the nineteenth century had been rapidly weakening prior to the depression, but it seems quite probable that the depression, as a result of a feeling of less security with regard to the future, has increased this trend and has tended to make rural people give more attention to enjoying the goods of life in the present rather than in the future. This is somewhat indicated by the very definite increase of interest in recreation and sociability which has been marked during the depression. The revival of home made amusements and the increased interest in developing more adequate facilities for satisfactory recreation within the community, has been a distinct gain to rural life, and is emphasizing the satisfactions of simple, informal, neighborly forms of play as distinctive rural values.

---

[2] For a brief bibliography on this general topic see Bercaw, Louise O. *Advantages and Disadvantages of Country Life*. Washington, D.C.: U. S. Department of Agriculture. Bureau of Agricultural Economics. Agricultural Economics Bibliography No. 37. May 1932. P. 30. Mimeo.

[3] *Cf.* Williams, James M. *Our Rural Heritage*. New York: Alfred A. Knopf. 1925. Chapter XI.

Whether their experience with inflated land values and of the advantages of a more stable agriculture, as shown by the relative security of debt-free farm owners during the depression, may mark a turning point from the type of farming which Warren H. Wilson[4] characterized as that of the "exploiter" and may have shown the more permanent values of the type which he called "the husbandman," remains to be determined, but seems quite possible. The general concern with the increase of tenancy and the values of farm ownership, as voiced, for instance, by the Farm Tenancy Commission of Arkansas[5] may be an indication of a beginning of a national policy which will seek to encourage stable farm ownership, and which has already been greatly helped by the Farm Credit Administration.

A definite appreciation of the values of the small farm for older people has arisen during the depression, has been strengthened by the experience of old age security administration, and has been recognized by debt adjustment committees working with the Farm Credit Administration.

The depression has undoubtedly sharpened the issue as to whether American agriculture should acquiesce in the trends described above by Dr. McCormick and accept the economic system and urban values incidental thereto, and the reduction of a large, perhaps the larger, part of rural folk to a condition commonly called peasantry; or whether it should recognize that there may be cultural values in a simpler form of rural life which will seek to maintain the independence of the mass of farmers, and the preservation of other values of family life, folk culture, and security, even at a relative loss of certain material advantages. This issue has been well stated by Dr. Carl C. Taylor in

---

[4] Wilson, Warren H. *The Evolution of the Country Community*. Boston: The Pilgrim Press. 1923. Revised ed. Chapter III

[5] *Findings and Recommendations of Hot Springs Meeting*. Hot Springs: Farm Tenancy Commission of Arkansas, appointed by Governor J. M. Futrell. C. E. Palmer, Chairman. Leaflet. November 24 1936. P. 6

his presidential address[6] before the American Country Life Association, in which he shows the possibility of a middle position between what he calls the "price and market rural culture" and a "peasant rural culture." He concludes:

I am inclined to the belief that we can, if we will, look forward to one that will combine the gains of commercial farming and the finer elements of a peasant culture. I am further inclined to the belief that if we strive for the first without striving for the second, we will gain the ultimate benefits of neither.

The same point of view was expressed in another form by Dr. H. C. Taylor, in his presidential address before the same association two years earlier, when he said:

The time has come when farmers should devote more time to acquiring that culture which our ancestors of necessity discarded in the pioneering days. It may be that instead of better farming and better business as a basis of better living, we must resort to less farming as a means to more living. Instead of throwing our entire energy into making dollars, we may well devote more of our time to the development of an individual culture, a family culture, and a community culture, all of which may enrich our country life far more than can be possible when the maximum of thought and energy is devoted to the dollar economy. . . .

. . . This improvement of the quality of life requires time for reading, time for meditation in quiet surroundings, time for social contacts without economic motive, time for the simple joys of being together; with literature, music, and art, as the media of common emotions and common aspirations. *Under such influences will develop ideas and attitudes with respect to human relations which make it seem more natural to give than to take, to protect than to injure, to love than to hate. People who have these qualities imbedded in their lives so that they practice them unconsciously and without motive or design are truly cultured.*[7]

The above discussion indicates the changes in rural culture and the redirection of ideals and appreciation of rural values

[6] Taylor, Carl C. "What Kind of Rural Life Can We Look Forward to in the United States?" *Country Life Programs. Proceedings of the 18th American Country Life Conference.* Chicago: University of Chicago Press. 1936

[7] Taylor, Henry C. "National Policies Affecting Country Life." *National Policies Affecting Rural Life. Proceedings of the 16th American Country Life Conference.* Chicago: University of Chicago Press. 1934. Pp. 26 ff.

which may have been affected by the depression, but it does not give us much clue as to what research may be necessary to validate the hypotheses formulated. The materials for determining the facts do not exist in any statistical form, and there is little documentary evidence beyond what reflection of public sentiment may be obtained from local newspapers and the agricultural press. As indicated in the previous section the best method will undoubtedly be to make surveys of representative sample areas by the use of attitude tests, which is an expensive method and therefore limits its general applicability.

Dr. Roy H. Holmes, of the University of Michigan, has developed an interesting method of obtaining rural opinion through a regular system of intimate correspondence with a well selected list of several hundred local leaders, which he expects to publish shortly.[8] The method demands a very large amount of time on the part of the investigator and adequate stenographic help, and there are questions with regard to the selection of the correspondents and the elimination of subjective interpretation of their replies which demand careful consideration, but the method, after Dr. Holmes has described and evaluated his procedure and its results, will perhaps reveal sufficient merit to warrant its further trial.

Possibly the collection of a large sample of personal correspondence between members of families and intimate friends which was written during the depression, and its organization and interpretation as was done by Thomas and Znaniecki in *The Polish Peasant,* might yield interesting results, but the amount of work involved makes it questionable whether the results would warrant it.

Whatever methods are used, it is more important than ever before that we obtain a better knowledge of the attitudes of rural people toward rural life and its distinctive values and

---

[8] See *Rural Sociology.* 2:59. March 1937

toward the changes which are occurring. In talking and corresponding with agricultural leaders the author has the distinct impression that in general they do not have as intimate a knowledge of farmers' opinions and attitudes as formerly, and that they have less concrete evidence upon which to base their impressions. This may be partly a subjective reaction, but there seems to be reason for this suspicion if one contrasts with former times the sort of contacts which persons outside the local rural community now have with farm people. Before the days of the automobile, speakers at farmers' institutes, which were then the chief form of what is now agricultural extension work, stayed overnight at farm homes and had ample opportunity for intimate conversation both in the home and as they drove back and forth in horse drawn vehicles. The same was true of the county agricultural agent, who spent more time with his constituents and came to know many of them intimately. Today the extension worker from the agricultural college, representatives of farmers' organizations, and the county agricultural agent all meet the farmers in group meetings which they reach by automobile from the county seat and then dash back again. The county agent, in addition, makes many calls a day, now and then stopping for dinner at some farm home, but he rarely stays overnight. There is less opportunity for conversation. The subtler attitudes of country folk can be sensed only by those who have frequent contact with them—new reason for developing definite systems of obtaining rural attitudes and public opinion upon some of these more intangible and yet fundamental values and objectives of rural life. Current events show, as never before, that we need to know accurately the real attitudes and feelings which dominate rural folk. Without this knowledge public policies may not take into account changes in rural attitudes which have undoubtedly been materially affected by depression experiences.

Chapter X

Concluding Remarks

In our survey of what has happened to the social aspects of rural life during the depression years, we have found little evidence of any occurrences which will necessarily permanently affect basic rural patterns. It is, however, too soon to get any adequate perspective of what has happened, and future historians may see, in the events since 1920, the beginning of tendencies which we cannot adequately evaluate at present. In general, the depression seems to have accelerated movements or trends already under way and has forced some of them upon public attention although, in some cases, as in the urban-rural population movement, there were temporary reversals of tendency which created a new situation.

Although we cannot yet discern fundamental changes in rural life brought about by the depression, it is clear that the depression has precipitated certain problems of adjustment whose solution will require the best research of social scientists. The general recognition of these problems and the determination to deal more effectively with them may be the most important contribution of the depression to future changes in rural life.

The revelation of the serious plight of the poorer agricultural classes, the tenants and croppers of the South, the migratory agricultural laborers and the farm families on large areas of marginal land, has forced the nation to consider how their condition may be permanently improved. The studies of the types of cases given relief have also shown the considerable number of rural families or individuals who are more or less permanently dependent and for whose care there has been wholly inadequate provision in the past. There have been various studies of these classes in the past, but mostly in terms of their function in the economic system of agriculture. When we are forced to consider how they may be made independent and self-supporting as an alternative to maintaining them indefinitely on public relief, we become aware that, under existing conditions, there is little possibility for most of them to improve their situation and that our boast of America's being the land of opportunity has little meaning for them. It is evident that much more thorough study is required of the factors which are responsible for rural poverty and which prevent its victims from the advancement that might seem possible. We need to know more of the sanitary conditions affecting health and the health facilities available, of the lack of educational opportunity and the failure to acquire an elementary education which will enable people to cope with the simplest problems of life, of the problems of overpopulation of certain areas, and of ways of reducing excessive fertility. These problems are by no means new, but only in a depression, because of the excessive cost of public relief necessary on account of neglect in the past, are we challenged to meet their solution.

Consideration of the position of the poorer agricultural classes during the depression raises the question as to whether a more definite stratification among different classes of farmers may not be taking place. Cleavage between the "haves" and the "have nots" always occurs during a depression, but the prospect for ameliorating this cleavage seems less promising than heretofore.

Because of prolonged unemployment in the cities the direction of migration from country to city, which had existed during this century, was temporarily reversed in 1932 and much slowed down in the other years from 1930 to 1935. Inasmuch as most of these migrants to cities had come from overpopulated regions, they tended to return to them and therefore created a serious problem of maintenance in areas least able to cope with it.

The depression, has, therefore, focused attention on problems of population as never before, particularly with regard to areas of excessive fertility and with regard to means of stimulating migration from areas where former means of livelihood no longer make it possible to maintain the existing population. Studies of mobility of rural population have assumed a new practical importance. Inasmuch as most of the rural-urban migration has been that of older youth, its stoppage has produced an excess of youth in many rural communities. Unless the tendency to reduce the number of employees in urban industries is reversed, it will be necessary to find vocational opportunities for these young people in rural communities, and rural institutions will be compelled to give their welfare a larger place in their programs. If but half or a third of the rural youth who have been migrating to cities in the past were to remain in the country, it would result in most important changes in the economic and social life of rural communities.

The partial stoppage of rural-urban migration, as well as the prospect for a larger number of permanently unemployed in the cities, raises the question as to whether there is a possibility of maintaining more people on the land with a reasonably satisfactory standard of living and acceptable social status. The considerable increase of part time farmers, especially near cities, and the larger proportion of small farms are some evidence that this movement is already under way. Students of Southern agriculture point out that there is a considerable amount of land here and there throughout the South, either idle or poorly worked, which might be made to yield a modest living if intensively farmed with emphasis on production for home use. Made more acute by the depression is the issue as to whether agricultural policy should be directed to the welfare of the commercial farmer, for maintaining the best living for the entrepreneurs who can produce the agricultural products necessary for the nation with the most efficiency; or whether it should also encourage the settlement of more people on the land who can produce much of their own living from it and raise enough products for sale to give them an income which, although it may not enable them to have all of the modern conveniences and comforts of life, may afford them a security and make possible a type of life which would have advantages over what they might enjoy as wage earners.[1] More precise knowledge of the factors involved challenge both agricultural economists and rural sociologists.

The period of the depression seems to have marked the definite turning of the individualistic attitudes of the mass of farmers to a belief in the necessity for collective action. Independent initiative was necessary for success in pioneer days, but when problems of marketing and business relations become more important than methods of production, the average farmer is helpless as an individual. Whether he is to achieve his ends through cooperative associations or other farmers' organizations, or through organization set up by the government, it is clear that agriculture recognizes the necessity of collective action as never before.

It is interesting to note, however, that although entrepreneur farmers organize to advance their own interests, they are like

---

[1] Thus in the conclusion of his study of the cotton plantation tenant Dr. T. J. Woofter makes the following statement: (Woofter, T. J. Jr. *Landlord and Tenant on the Cotton Plantation*. Research Monograph V. P. 183. Washington, D.C.: Works Progress Administration, Division of Social Research. 1936)

"It is not outside the realm of possibility that the two trends may develop simultaneously with larger and more mechanized farms for the production of money crops, interspersed with increasing numbers of small family-sized farms which are operating primarily to produce a living for a family. These smaller farms must include acreages of money crops to pay for taxes, clothing and services, but in a region of as varied potentiality as the South they can also produce the major portion of the family food requirements." He also says (p. 181): "Nor do the data in this study support the theory that to remain and endeavor to work out a different agricultural economy from the one at present practiced would necessarily mean greater poverty than would be the case if the migrants join the millions who exist on the margin of the industrial labor market."

manufacturers and business men in denying the right of their employees to do the same; witness the violence with which they seek to suppress the organization of migratory laborers in California and the organization of the croppers in Arkansas.

This decline in the independence of farmers is also revealed in their increasing use and demand for public services supported by government, such as education, health, extension and welfare services. Like the city wage earners they are unable to support these services by direct contributions or wholly by direct taxation, so they demand county, state, and national appropriations for their support, as a means of giving them their share of advantages through equalizing the cost and thus redistributing the general income through taxation.

These are among the more important changes, occurring during the depression, which have brought about a new definition of the rural situation, both in the attitudes of rural people and in the public opinion of all classes. As a result of these changes in our understanding of the problems which must be met to produce a more satisfactory rural society, we have seen that there are many problems concerning which we lack adequate knowledge and about which extensive research is necessary to obtain the essential facts. It is not necessary to review or sumarize the various lines of research suggested throughout this report, but it may be useful to call attention to certain types and methods of research which need emphasis in view of their frequent recurrence in the analysis of the problems raised.

It is evident that in the analysis of many problems, particularly those which involve the use of census data such as population, fertility, tenancy, etc., national and state averages and trends often have little meaning; and that it is desirable to break data down to a county basis and to reassemble them by crop areas, agricultural sections, and cultural areas or regions. Only by such procedure can the influence of diverse factors within the purely political boundaries of states be revealed, and can associations be made which will be significant in revealing causal relationships. We cannot describe the conditions of rural life for this country as a whole, or even for states as a whole, for within these areas there are fundamental differences for different sections. Successful public policy cannot be uniform, but must be so framed as to be used in those areas to which specific policies are applicable as shown by a clear delimitation of their characteristics.

One of the chief difficulties in determining what changes have occurred in rural life during the depression is in the lack of data concerning previous conditions over a sufficient period to indicate trends. If we are to determine the social effects of periods of depression or prosperity, we must have recurrent studies to reveal the differences. The value of such recurrent studies is best illustrated by the surveys of 140 representative village community areas made under the direction of Dr. E. deS. Brunner in 1924, 1930, and 1936. This is the only body of knowledge concerning rural social conditions which covers the whole country for three dates except that in the federal census, and has well demonstrated its value. It should be continued on a larger scale to include a larger number of villages of the same type, so as to provide for dropping out those which are no longer rural, and should include an equal number of villages of 500 or less population, both incorporated and unincorporated. The only other studies which have been repeated are those of Lively on mobility in Ohio, of Kirkpatrick on the standard of living in Wisconsin, and of Von Tungeln of one rural community in Iowa. On all of these topics it would be desirable to have surveys repeated at definite intervals in a sufficient number of cases to make the sample as representative as possible.

Although, in last analysis, all scientific generalizations must rest upon a quantitative measurement of frequency of occurrence of given phenomena, the initial discovery of relationships and the development of working hypotheses are by no means dependent upon quantitative measurements. Thus, for a better understanding of the type of families who live on marginal land, of those who are on relief and may be rehabilitated, and for determining the effect of type of farming on family relationships, it is desirable to have a considerable body of case studies which will be as much concerned with a clinical analysis of each case as with gathering a schedule of facts as a basis for statistical analysis. Such case studies require a superior type of training upon the part of the field worker (who must have more of the qualifications of the social case worker), and they require much more time than merely taking a schedule; but it is probable that, through this method, insights with regard to social relationships would be obtained which could not otherwise be discovered.

It has also been shown that one of the most important aspects of change in the rural social situation consists of whatever changes may have occurred in the attitudes of rural people, with regard to those values which they hold most essential and which motivate their behavior. In this field of the study of the social attitudes of rural people little has been done, but the studies which have been made reveal the possibilities. With the rapid change in ideologies which is now taking place among rural as well as urban people, and with the reconsideration of old values and the necessity for decision with regard to new values, it is of prime importance that means be devised for a continual appraisal of rural attitudes and the development of adequate and practicable techniques for this purpose. Both intensive and extensive study of rural attitudes is necessary if we are to have an adequate understanding of the social psychology of rural people as a basis of procedure in whatever policies may be attempted for the improvement of rural life.

The length of the agricultural depression and the sudden reversal from the prosperity of the first two decades of this century brought about an intensive scientific study of the economic problems of agriculture which has demonstrated its utility. If the problems of the social aspects of rural life which have been made acute during the industrial depression have revealed the importance of obtaining a more accurate and adequate knowledge concerning them, through scientific methods of research, and if this situation gives rise to the means for carrying on such research, both under governmental and private auspices, it is probable that a more realistic view of the problems of rural life may be obtained and that facts may be made available which will implement measures for rural improvement.

Index

Adams, James Truslow, 6n
Adams, R. L. and Wann, J. L., 41n
Adult education, 108 ff.
Agricultural Adjustment Act, 109
Agricultural Adjustment Administration, reason for, 2, 142; and mobility of cotton tenants and croppers, 56; differential benefits to entrepreneurs and tenants, 63; farmers' support of, 110; county program planning, 112; relation to county agents, 114; dependence of farmers on, 129
Agricultural depression, period of, 3n; extent, 4; see also Depression
Agricultural Economics, U. S. Bureau of Estimates: of farms changing ownership, 5; of urban-rural and rural-urban migration, 14, 19; increase in number of farms and persons on farms, 17; studies of rural mobility, 21; and resurvey of 140 villages, 23, 156; study of: resettlement, 50; tenancy, 58; farm labor problems, 61; standard of living on farms, 125; Carl C. Taylor, quoted, 144
Agricultural Experiment Station, Kentucky, 37
Agricultural extension work, 107, 108 ff
Agricultural readjustment problems, social corollaries; regionalism, 29-52; social effects of marginal land, 35; part time farming, 38; rehabilitation and resettlement, 45 ff.; federal agencies for relief, 2; changing markets: expansion ended, 6; social aspects of readjustment, 29-52; extension work, 107, 108 ff.; attitudes toward the future of, 123-40; changes in standard of living in, 124-29; possibility of government control, 129; membership in farmers' organizations, 132 ff.; tradition of opportunity, 139; planned economy movement, 143; consideration of poorer classes a depression result, 151; see also Farmers: Farms
Agriculture, U. S. Department of, 76, 112
Alabama, tenancy, 55
Alexander, W. W., see Johnson, Charles S., Embree, Edwin R., and Alexander, W. W.
Allen, R. H., see Troxell, W. W., et al
Allin, Bushrod W., see Goodrich, Carter, Allin, Bushrod W. and Hayes, Marion
Allin, Bushrod W. and Parsons, Kenneth H., 17, 18
American Farm Bureau Federation, membership, 132 ff.
American Federation of Labor, 135
Anderson, C. Arnold and Smith, T. Lynn, 9n
Anderson, W. A., 24, 68n, 71; and Kerns, Willis, 68n, 71n
Appalachian-Ozark area, 30, 36; migration in, 22, 23, 24, 26; organization in, 34; part time farming in, 42; dependent class in, 64
Arizona, mobility in, 21; drought effect in, 60

159

4 ff.; frustration reactions, 7; population changes, 11-28; depression readjustment problems, 29-52; affected by types of farming, 33; farmers' social and economic status, 53-67; cotton acreage reduction program effect on, 59, 63; small farm in, 61 ff.; youth problems, 68-78; depression effects on institutions of, 79-102; services affected by depression, 103-22; farmer attitudes toward future, 123-40; rural-urban relationships, 141 ff.; distinctive attitudes and culture influenced by depression, 141-50

Rural teachers, 85

Ruritan Clubs, 99

Salter, L. A., Jr., and Darling, H. D., 39n; see also Davis, I. G. and Salter, L. A.

School, rural, depression effects on, 83 ff.; increased attendance at, 84; public attitude toward, 84, 90; consolidation, 85 ff.

Scope and Method Series, Black, John D., ed., 9

Security, farmers' feeling of, 136 ff.; farms for old people, 147

Sedgwick County, Kansas, church membership survey, 93

Services, 103-22; health services and facilities, 103 ff.; recreation and play, 106 ff.; extension work and adult education, 108 ff.; social welfare work, 115 ff.

Short-grass areas, 30, 36

Simkin, Francis B., 63n

Sitterley, J. H., see Morison, F. L. and Sitterley, J. H.

Smith, F. V. and Lloyd, O. G., 41n

Smith, T. Lynn, 98; see also Anderson, C. Arnold and Smith, T. Lynn; and

Frey, Fred C. and Smith, T. Lynn

Smith-Hughes Act, 71

Sneed, Melvin W. and Ensminger, Douglas, 92n, 95

Social aspects of agricultural readjustment, 29-52

Social Science Research Council, 32, 50

Social Security Act, 47, 82

Social welfare, see Welfare work

. . . *Social Work in the Depression,* 48n, 115n

South, the, organization in, 34; tenancy in, 54 ff.; economic stratification in, 63; birth rate in, 80; housing, 81; illiteracy in, 88; county-wide community life, 100; rural standards of living, 127

South Carolina, population trends, 18

South Central States area, size of farms, 138

South Dakota, mobility in, 21

Southern proletariat, 63 f.

Southwest area, migratory labor in, 59

Spencer, Lyle M., see Stouffer, Samuel A. and Spencer, Lyle M.

Standard of living, depression-effected changes, 124 ff.

State Relief Administration, California, 60

Steiner, Jesse F., . . . *Recreation in the Depression,* 75n, 106n

Stouffer, Samuel A., vi, 14n; and Lazarsfeld, Paul F. . . . *Family in the Depression,* 13n, 44n, 70n, 79n; and Spencer, Lyle M., 79n

Stratification and status of farmers, 53-67

Submarginal land, problem of, 26, 35 ff.

Subsistence Homesteads Division, 39

Surveys, FERA studies, 21, 46, 80, 115, 121; regional, 32; relation of marginal lands to living conditions, 36;

Studies in the Social Aspects
of the Depression

AN ARNO PRESS/NEW YORK TIMES COLLECTION

Chapin, F. Stuart and Stuart A. Queen.
Research Memorandum on Social Work in the Depression. 1937.

Collins, Selwyn D. and Clark Tibbitts.
**Research Memorandum on Social Aspects of Health in the Depression.
1937.**

The Educational Policies Commission.
Research Memorandum on Education in the Depression. 1937.

Kincheloe, Samuel C.
Research Memorandum on Religion in the Depression. 1937.

Sanderson, Dwight.
Research Memorandum on Rural Life in the Depression. 1937.

Sellin, Thorsten.
Research Memorandum on Crime in the Depression. 1937.

Steiner, Jesse F.
Research Memorandum on Recreation in the Depression. 1937.

Stouffer, Samuel A. and Paul F. Lazarsfeld.
Research Memorandum on the Family in the Depression. 1937.

Thompson, Warren S.
Research Memorandum on Internal Migration in the Depression. 1937.

Vaile, Roland S.
**Research Memorandum on Social Aspects of Consumption in the
Depression. 1937.**

Waples, Douglas.
**Research Memorandum on Social Aspects of Reading in the Depression.
1937.**

White, R. Clyde and Mary K. White.
**Research Memorandum on Social Aspects of Relief Policies in the
Depression. 1937.**

Young, Donald.
Research Memorandum on Minority Peoples in the Depression. 1937.